ALBATROSS
Elusive Mariners of the Southern Ocean

ALBATROSS
Elusive Mariners of the Southern Ocean

ALEKS TERAUDS
& FIONA STEWART

First published in Australia in 2005 by Reed New Holland
an imprint of New Holland Publishers (Australia) Pty Ltd

SYDNEY • AUCKLAND • LONDON • CAPE TOWN

14 Aquatic Drive Frenchs Forest NSW 2086 Australia
218 Lake Road Northcote Auckland New Zealand
86 Edgware Road London W2 2EA United Kingdom
80 McKenzie Street Cape Town 8001 South Africa

Copyright © 2005 in text and photographs: Aleks Terauds except where credited otherwise
Copyright © 2005 in illustrations: Fiona Stewart except where credited otherwise
Copyright © 2005 in maps: as credited
Copyright © 2005 New Holland Publishers (Australia) Pty Ltd

All rights reserved. No part of this publication may be reproduced, stored in a retrieval system or transmitted, in any form or by any means, electronic, mechanical, photocopying, recording or otherwise, without the prior written permission of the publishers and copyright holders.

NATIONAL LIBRARY OF AUSTRALIA CATALOGUING-IN-PUBLICATION DATA:

Terauds, Aleks, 1971- .

    Albatross: Elusive Mariners of the Southern Ocean.

    Bibliography.
    ISBN 1 877069 26 4.

    1. Albatrosses. 2. Albatrosses - Breeding - Australia.
    I. Stewart, Fiona, 1976-. II. Title.

    598.42

PUBLISHER: Louise Egerton
PROJECT MANAGER: Yani Silvana
DESIGNER: Fiona Stewart
REPRODUCTION: Photolith Reprographics
PRODUCTION: Linda Bottari
PRINTER: Toppan Printing Company, China

TITLE PAGE IMAGE: Short-tailed Albatross (*Diomedea brachyura* Temm.), 1884.
Watercolour by H. C. Richter in Gould's *The Birds of Australia*. National Library of Australia.
COVER IMAGE: Black-browed Albatross, Macquarie Island. Photo: Aleks Terauds.

Contents

Acknowledgements	7
Introduction	9
Breeding Areas	19
Macquarie Island	23
Albatross Island	35
Pedra Branca	41
The Mewstone	47
Humans in Bass Strait and the Southern Ocean	53
Whaling	55
Sealing—Tasmania	57
Sealing—Macquarie Island	59
Seabirds	63
Feral Animals—The Sealers' Legacy	64
Pelagic Tuna Fisheries	66
Patagonian Toothfish Fisheries	70
Trawling	71
The Albatrosses	75
Wandering Albatross	79
Shy Albatross	93
Black-browed Albatross	105
Grey-headed Albatross	119
Light-mantled Sooty Albatross	133
Conservation	145
Tasmanian Islands	146
Macquarie Island	150
Living and Working on the Islands	155
Albatross Island	155
Pedra Branca	156
The Mewstone	156
Macquarie Island	157
References	163
Artists and Photographers	171
Index	173

A perfectly groomed pair of Black-browed Albatrosses stand side by side on the southern slopes of Petrel Peak, Macquarie Island.

Acknowledgements

THANKS FIRST TO NIGEL BROTHERS (formerly of Parks and Wildlife Service Tasmania and Nature Conservation Branch, Department of Primary Industries, Water and Environment [DPIWE]) and Rosemary Gales (Nature Conservation Branch).

Nigel pioneered research into the Shy Albatrosses on Albatross Island over two decades ago and both he and Rosemary initiated the long-term monitoring program on Macquarie Island. Their dedication to the conservation of albatrosses is inspirational and they have contributed significantly to reducing the number of albatrosses that are caught and killed on long-line fishing vessels around the world.

April Hedd (University of Tasmania), Sheryl Hamilton and Rachael Alderman (Nature Conservation Branch) also contributed much to our understanding of Shy Albatrosses and have obtained information that was extremely useful in assisting the conservation of this species. Travel to the Tasmanian islands would not have been possible without the assistance of Neil Smith (*Wild Wind*) and the Tasmanian Marine Police. Travel to Macquarie Island was usually facilitated by the Australian Antarctic Division.

The Albatross Project on Macquarie Island has succeeded through the efforts of many people. Geof Copson, Jenny Scott, Karen Alexander, Al Wiltshire, Sam Thalmann, Fiona Hume, Jason Hamill, Rachael Alderman, Jacinda Amey, Martin Schulz, Sue Robinson and Ken Smith all worked in the field on Macquarie Island and without their dedication and commitment in an often challenging environment much of the work would not have been possible. Thanks also to the other expeditioners and members of the Australian National Antarctic Research Expedition (ANARE) who made the trips down south possible and provided assistance in the field. Special thanks to Justine Shaw, Mark Geyle and Robb Clifton for their assistance in this.

The assistance of many other people was essential to the maintenance of these long-term programs: Alastair Richardson, Mark Hindell, Kit Williams (University of Tasmania) provided valuable advice, input and technical expertise. Barry Baker, Belinda Dettmann, Lisa Hardy and David Drynan of the Department of Environment and Heritage also provided much useful advice and support.

The programs have been funded by Nature Conservation Branch, DPIWE; Environment Australia (now DEH); the Australian Antarctic Division, the Antarctic Scientific Advisory Committee (ASAC); the University of Tasmania; the Australian Research Council; the MA Ingram Trust; and the Royal Zoological Society of NSW. Special thanks must go to the Nature Conservation Branch, DPIWE Tasmania for the ongoing support provided to both programs.

Thanks also to Robert Morrison for his guidance with design and Mike Sumner for advice on creating the maps. This book would not have been possible without New Holland Publishers, and special thanks to Louise Egerton and Yani Silvana for their advice and support.

A six-month-old Wandering Albatross chick anxiously awaits the return of its parents on a snow-covered nest on Petrel Peak. Mt Haswell and Caroline Cove Amphitheatre are in the background.

Bathed in the early morning light of sunrise these three-month-old Shy Albatross chicks will lose their fluffy down to a sleeker adult coat in only a few weeks.

INTRODUCTION

ALBATROSSES ARE THE LARGEST MEMBERS of the petrel family (order Procellariiforme) and are comprised of 24 different species in four main genera.[1] This book describes five species that breed in Australian waters: four on subantarctic Macquarie Island and one on offshore islands around Tasmania.

There are over 8000 pairs of Wandering Albatrosses (*Diomedea exulans*) breeding worldwide[2] but only 19 of these breed on Macquarie Island. Black-browed (*Thalassarche melanophrys*) and Grey-headed Albatrosses (*T. chrysostoma*) also have very small breeding populations on Macquarie Island (they make up less than 1 per cent of the global population); Light-mantled Sooty Albatrosses (*Phoebetria palpebrata*) are more common with over 1000 breeding pairs on Macquarie Island, constituting around 5–10 per cent of the global population.[2] The Shy Albatross (*Thalassarche cauta*) only breeds in three main colonies around Tasmania: Albatross Island, home to approximately 5000 nesting pairs each year, Pedra Branca, a tiny rocky island that sustains approximately 200 breeding pairs each year, and the Mewstone, another impressive rock island where over 7000 breeding pairs of albatrosses live and breed each year.[3] The latter two islands are off the south coast of Tasmania and are wild, windswept places on the northern edge of the Southern Ocean.

Albatrosses are long-lived seabirds that are characterised by a late onset of sexual maturity, single-egg clutches, low breeding success, and naturally high adult survivorship.[4] They are extremely long lived and spend most of their lives travelling the ocean. Few species (terrestrial or marine) exhibit such extreme life history attributes and the evolution of such traits is likely to be a response to maximising lifetime reproductive success in the harsh but stable marine and terrestrial environments that these seabirds inhabit.[5] Because of these traits, albatrosses, at both population and species levels, are exceptionally vulnerable to 'unnaturally' high mortality rates[6] and recovery of populations is slow, particularly if the threat is ongoing.

Albatrosses typically breed in the higher latitudes of the world's hemispheres and, with the exception of Waved (*Phoebastria irrorata*), Black-footed (*P. nigripes*) and Laysan Albatrosses (*P. immutabilis*), all inhabit oceanic islands in temperate regions. The number and intensity of studies on albatrosses have increased since the late 1970s and 1980s when the first signs of declining populations in the Southern Ocean were identified. Declines in Wandering Albatross populations were first reported on South

A six-year-old male Wandering Albatross displays the profile of his large bill. The overhang of the top mandible found in all albatrosses helps them to hold their prey.

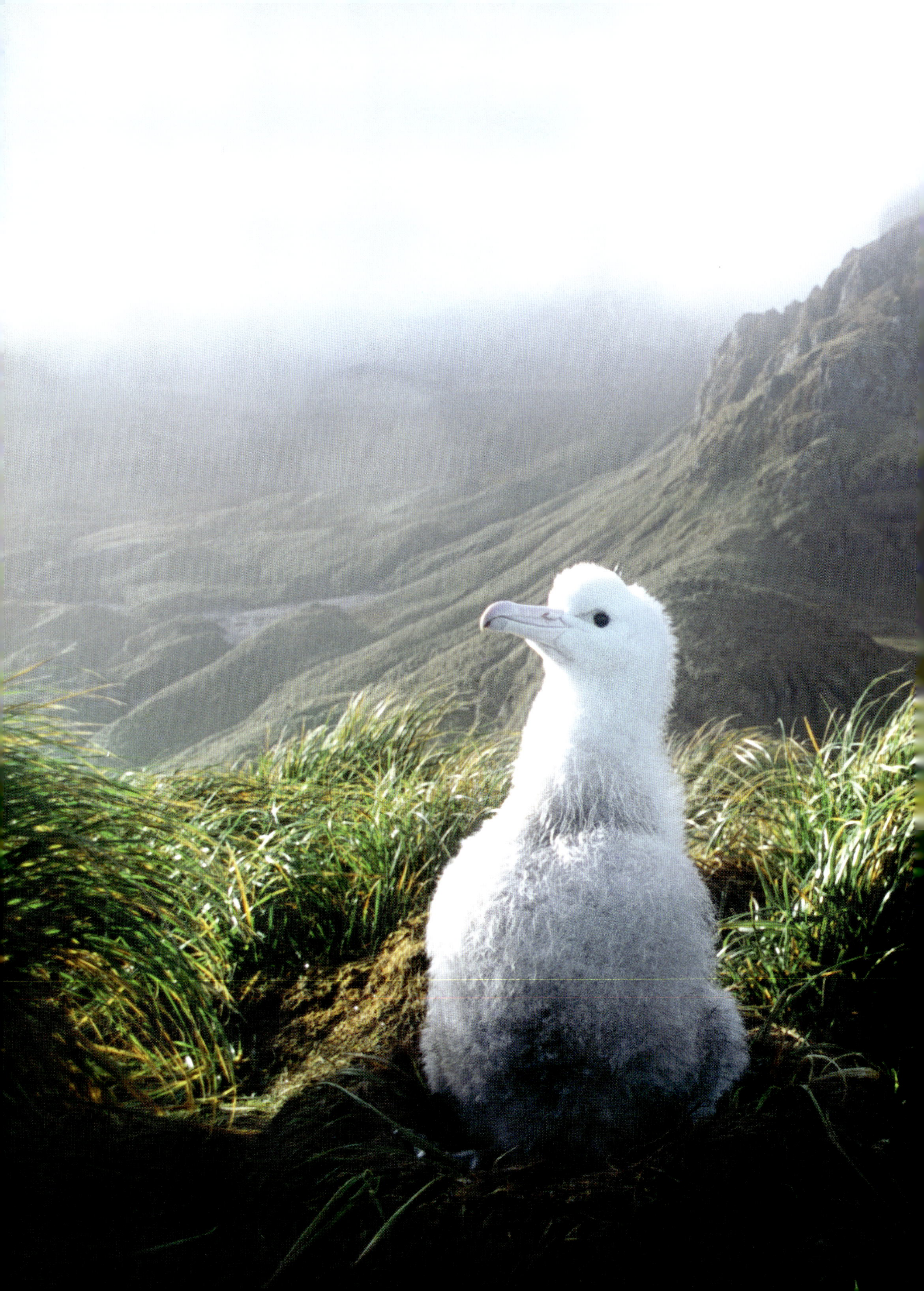

Among the lush, grassy tussocks on Petrel Peak, Macquarie Island, this four-week-old Wandering Albatross chick has been left alone by its parents while they forage for food.

Georgia,[7] Macquarie Island[8] and Iles Crozet.[9,10] It was the reports from Iles Crozet that first linked the observed declines to increased mortality due to long-line fisheries. The links between albatross deaths and driftnet fisheries in the northern hemisphere had been documented as early as the 1960s with reports of Black-footed, Laysan and Short-tailed Albatrosses (*P. albatrus*) all being caught and killed in large numbers throughout the 1960s and 1970s.[11]

More recently, studies have conclusively documented the decline of albatross populations throughout the southern hemisphere. Breeding locations at which these declines have been recorded include South Georgia,[12] Iles Crozet and Iles Kerguelen,[13] Amsterdam Island,[14] Marion Island,[15] New Zealand[16] and Macquarie Island.[17] In all of these studies, the declines were associated with increased mortality due to interactions of the albatrosses with long-line fisheries.

Long-line fleets fishing for Tuna (*Thunnus* spp.) were widespread throughout temperate latitudes of the southern hemisphere throughout the 1960s, 1970s and 1980s.[18] The potential scale of the impact of tuna long-line fisheries on albatrosses was highlighted by the work of Nigel Brothers, whilst working for the Parks and Wildlife Service, Tasmania, in the late 1980s and early 1990s,[19] and subsequent studies provided more data on the extent and nature of the interactions. In two comprehensive reviews of the conservation status and threats to albatross populations worldwide, Rosemary Gales, a Tasmanian marine biologist, described long-line fishing as the most serious threat to the survival of albatross populations.[2,6] The relatively recent

Effortlessly, this Grey-headed Albatross soars over West Rock, Macquarie Island.

advent of a large-scale fishery for Patagonian Toothfish (*Dissostichus eleginoides*) in Southern Ocean waters also presents a significant threat to albatrosses foraging in some areas.[20]

All five species of the albatrosses featured in this book have suffered from the activities of humans, from feather collecting on Albatross Island in the 1800s to the countless thousands of birds that have been caught and killed on longline fishing vessels from the 1960s to the present day. Three of these species—the Wandering Albatrosses, Black-browed Albatrosses and Grey-headed Albatrosses—are particularly at risk due to the small size of their breeding populations on Macquarie Island.

Wandering Albatrosses are the largest of the five species and typically show a biennial breeding pattern after a successful breeding attempt. That is, once the chick has left the nest (fledged), the parents take a year off before returning to breed in the following season. Most Grey-headed and Light-mantled Sooty Albatrosses also tend to breed every second year following a successful breeding attempt. Black-browed Albatrosses and Shy Albatrosses generally breed every year regardless of the outcome of the previous breeding attempt.

The timing of the breeding cycle also differs between species. Black-browed Albatrosses lay from late September to the end of October and chicks fledge in late March and April. Grey-headed Albatrosses generally lay through October and chicks fledge in April and May. Light-mantled Sooty Albatrosses lay in late October through to early November and chicks fledge through May and early June. Shy Albatrosses have a slightly longer breeding cycle with breeding birds beginning to arrive back on the colony throughout late June and July. Most eggs are laid in August and chicks hatch in late November, early December and start to fledge in April and May. Wandering Albatrosses, which are larger, have a significantly longer breeding cycle, with eggs laid in December and chicks fledging in November or December of the following year.

Studies of the Shy Albatross on Albatross Island have been conducted for over two decades. Largely pioneered by Nigel Brothers in the early 1980s and continued by Rosemary Gales, April Hedd, Sheryl Hamilton and Rachael Alderman in more recent times, these long-term studies have provided a wealth of knowledge on this species, which only breeds in Tasmanian waters. Aspects studied over this time include population demographics, foraging ecology (via satellite-tracking) and breeding biology.

ABOVE: A young female non-breeding Wandering Albatross rests on the summit of Petrel Peak.

LEFT: This pair of Grey-headed Albatrosses is prospecting for a suitable nesting site on the southern slopes of Petrel Peak.

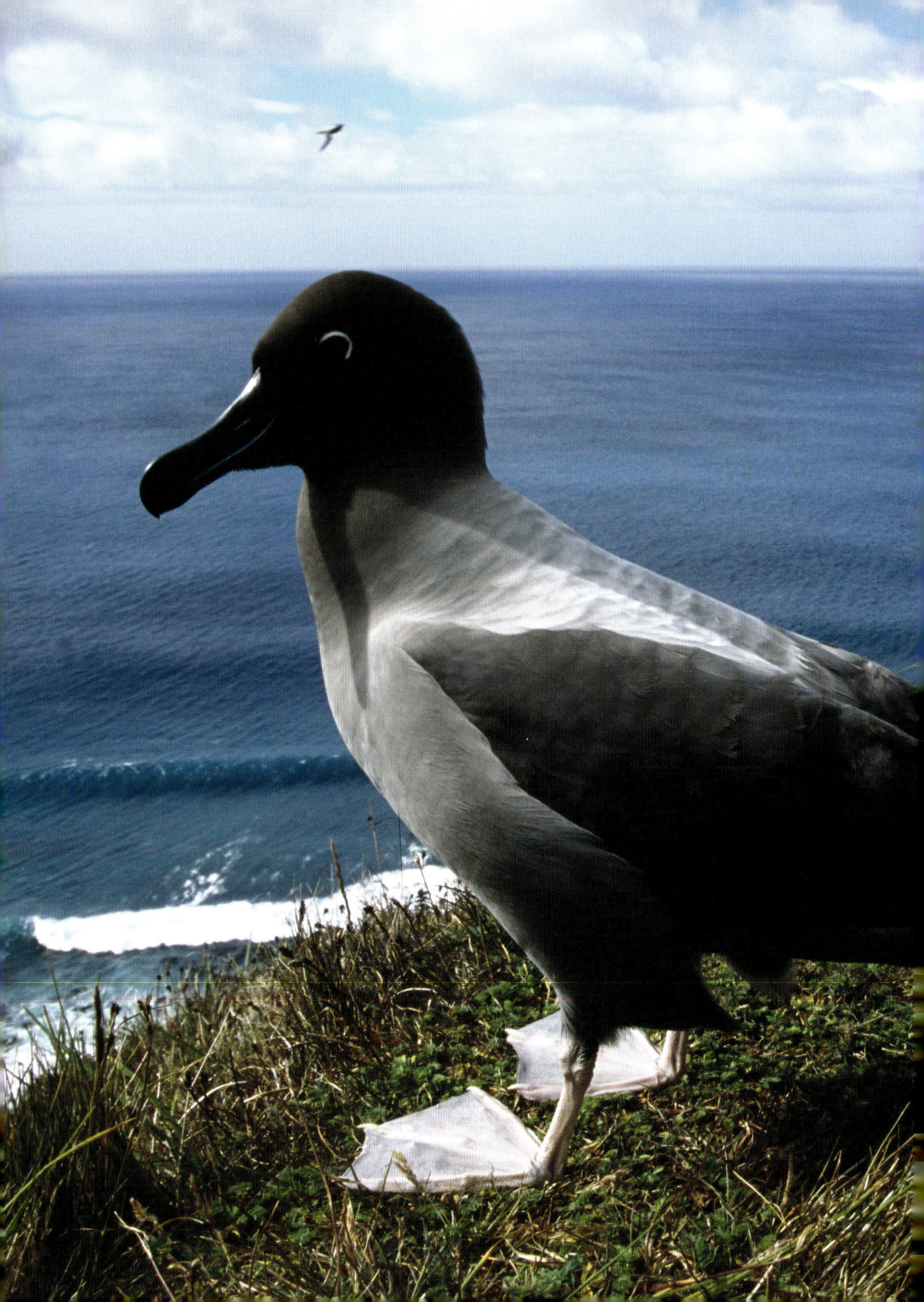

A pair of Light-mantled Sooty Albatrosses search together for a suitable nesting site along the slopes of Lusitania Bay, Macquarie Island.

Non-breeding Shy Albatrosses congregate in the Main Colony of Albatross Island. The Main Colony is one of four breeding colonies on the island and stretches for about 250 metres from north to south.

The declines reported in Southern Ocean albatross populations, and the increasing concerns over the number of albatrosses killed as a result of long-line fisheries, prompted the initiation of a comprehensive long-term study of the population trends, breeding biology and foraging ecology of the albatrosses on Macquarie Island in the early 1990s. In addition to concerns about high mortality, the study was also designed to complement albatross research in a global context. The population status of many breeding populations worldwide was determined throughout the 1990s[2] and the inclusion of the little-studied Macquarie Island breeding populations in this global pool of data was considered critical, particularly in light of the small and vulnerable breeding populations.

Due to the impact of past and/or present fishing practices, or their restricted breeding range, all these species are listed as *Vulnerable* or *Near Threatened* on a global scale (International Union for Conservation of Nature [IUCN] listing) and as *Endangered* or *Vulnerable* under Tasmanian legislation (see table below). The efforts of many people have helped to reduce the number of birds that are caught and killed in many areas of the world. However, despite these efforts many populations are still in decline. While interactions with fishing vessels continue to be a major cause of death away from their breeding islands, other human activities are also likely to negatively impact upon these birds. Climate change, marine debris, pollution, changes in food availability and feral animals in their breeding areas all potentially contribute to declines of albatrosses around the world and ongoing conservation efforts are essential to help minimise the destructive impacts wrought by human beings on these remarkable species of seabird.

The dangling feet of the Light-mantled Sooty Albatross help it to slow down as it comes in to land on North Head, Macquarie Island.

NUMBER OF BREEDING PAIRS AND CONSERVATION STATUS LISTING

SPECIES	NUMBER OF ANNUAL BREEDING PAIRS		POPULATION STATUS LISTING		
	Australia	*Global*	*IUCN*	*National*	*Stat*
Wandering Albatross	7–15	8500	Vulnerable	Vulnerable	Endangered
Black-browed Albatross	38–45	680 000	Endangered	Protected	Vulnerable
Grey-headed Albatross	65–80	92 300	Vulnerable	Vulnerable	Vulnerable
Light-mantled Sooty Albatross	~1100	21 600	Near Threatened	Not listed	Vulnerable
Shy Albatross	12	12 000	Near Threatened	Vulnerable	Vulnerable

Source: 'The recovery plan for albatrosses and giant petrels', Environment Australia, October 2001 and IUCN.

The map shows the Southern Ocean breeding sites and the Antarctic Circumpolar Current.

AUSTRALIA

NEW ZEALAND

◉ Antipodes Islands
◉ Auckland Islands
◉ Campbell Island
◉ Pedra Branca
◉ The Mewstone
◉ Albatross Island
◉ Macquarie Island

ANTARCTICA

◉ Heard Island
◉ Iles Kerguelen
◉ Iles Crozet
◉ Prince Edward Islands

Antarctic Circumpolar Current and associated fronts

◉ South Georgia
◉ Falkland Islands
◉ Chilean offshore islands

◉ ALBATROSS BREEDING SITES

0   1000 km

BREEDING AREAS

Most albatrosses breed in the southern temperate areas of the world. Only four species breed in the northern hemisphere. The Waved Albatross breeds only on the Galapagos Islands, while the Black-footed and Laysan Albatrosses breed on small Hawaiian islands in large numbers.

The only other species to breed in the northern hemisphere is the Short-tailed Albatross, which is endemic to temperate offshore islands around Japan. The southern temperate regions of the world provide an ideal habitat for these seabirds, which have evolved to be at home in the oceanic environment.

In the southern hemisphere there is a large ocean current system that encircles Antarctica, known as the Antarctic Circumpolar Current (ACC) (see map opposite). Here, warmer waters from northern regions meet colder waters from Antarctica. This current and mixing of water masses not only influences the global climate;[1] it also concentrates food in areas known as 'fronts'. It is no coincidence therefore that many albatrosses live and breed on the subantarctic islands situated close to or within the Antarctic Circumpolar Current, allowing them easy access to these productive waters. As well as these productive areas associated with the Circumpolar Current, albatrosses also often feed in waters close to the edge of the continental shelves, where deeper and shallower waters mix, creating upwellings that are high in nutrients and food.

Of the five species described in this book, only the Shy Albatross breeds close to Tasmania—on three islands: Albatross Island, Pedra Branca and the Mewstone (right). Albatross Island lies off the north-west corner of Tasmania, and Pedra Branca and the Mewstone lie off the south coast. All three islands are relatively close to the edge of the continental shelf, where there are productive waters in which the albatrosses are more likely to find food. The Wandering, Black-browed, Grey-headed and Light-mantled Sooty Albatrosses all breed on Macquarie Island, approximately halfway between Australia and Antarctica (see map opposite), but all of them also breed at other locations. The Wandering Albatrosses, for example, also breed on Marion Island and South Georgia, while Black-browed Albatrosses breed right around the southern hemisphere, including Heard Island, Iles Kerguelen, the Falkland Islands and on Chilean offshore islands.

The table on page 21 summarises the main breeding sites for Wandering, Shy, Black-browed, Grey-headed and Light-mantled Sooty Albatrosses and the locations of these breeding

The red dots indicate the three Shy Albatross breeding sites in Tasmanian waters.

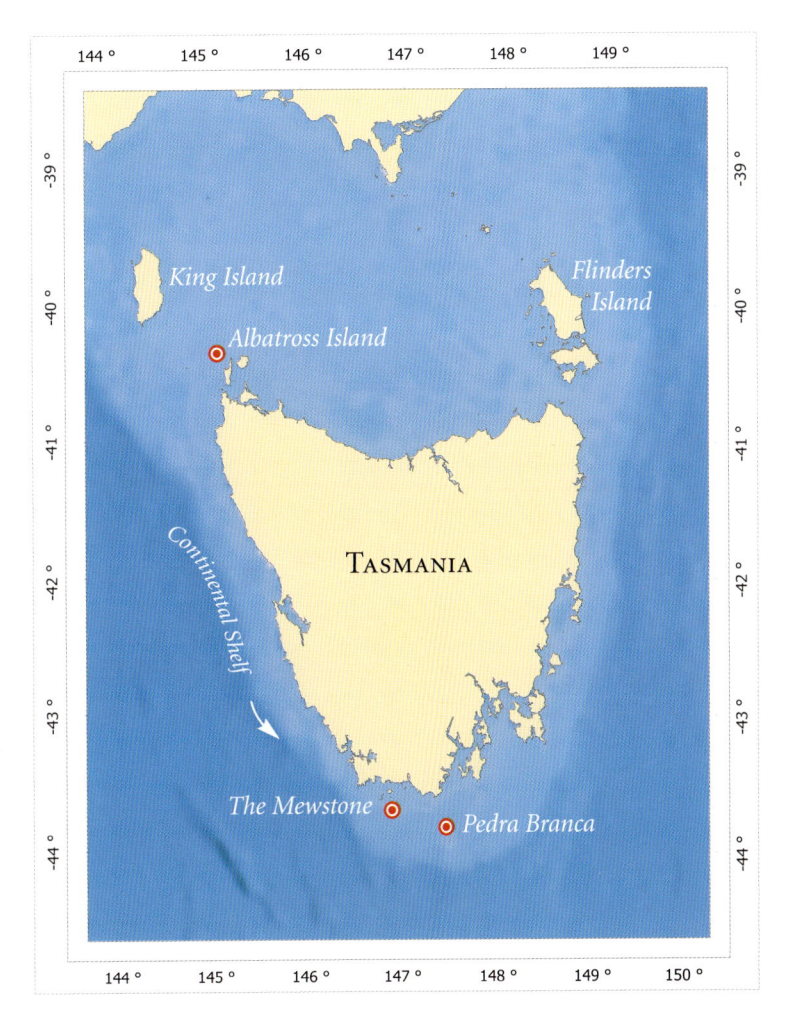

sites are also shown on the map on page 18.

Where albatrosses breed and forage is largely determined by where their food is to be found. Albatrosses spend most of their time at sea and all that they eat is obtained from the ocean. They actively hunt a variety of food, including squid, small fish and/or krill, but they are also scavengers, occasionally feeding opportunistically on ocean carrion or discards from fishing vessels. They are well adapted for life at sea, with the ability to fly long distances in search of food and find it with the assistance of an incredibly sensitive sense of smell. The food of albatrosses tends to be patchily distributed so they typically focus their foraging efforts in regions where this food is concentrated.[2] They target productive waters, particularly where water masses of different temperatures meet or where localised currents and eddies form.[3,4] Albatrosses may feed in relatively shallow waters within hundreds of kilometres of their breeding site or they may undertake massive oceanic foraging voyages, covering thousands of kilometres in a single trip.[5]

Prior to the advent of appropriately sized satellite-tracking devices, knowledge of specific foraging locations was primarily obtained through the recoveries of banded birds or at-sea observations. Since the first Wandering Albatrosses were tracked on Iles Crozet in 1989,[6] satellite-tracking studies have proven to be an extremely effective method of investigating the foraging trips of many species of albatrosses and identifying the areas in which they feed.

The red dot indicates the location of Macquarie Island in the Southern Ocean on the Macquarie Ridge. Also shown are the fronts associated with the Antarctic Circumpolar Current.[11,25]

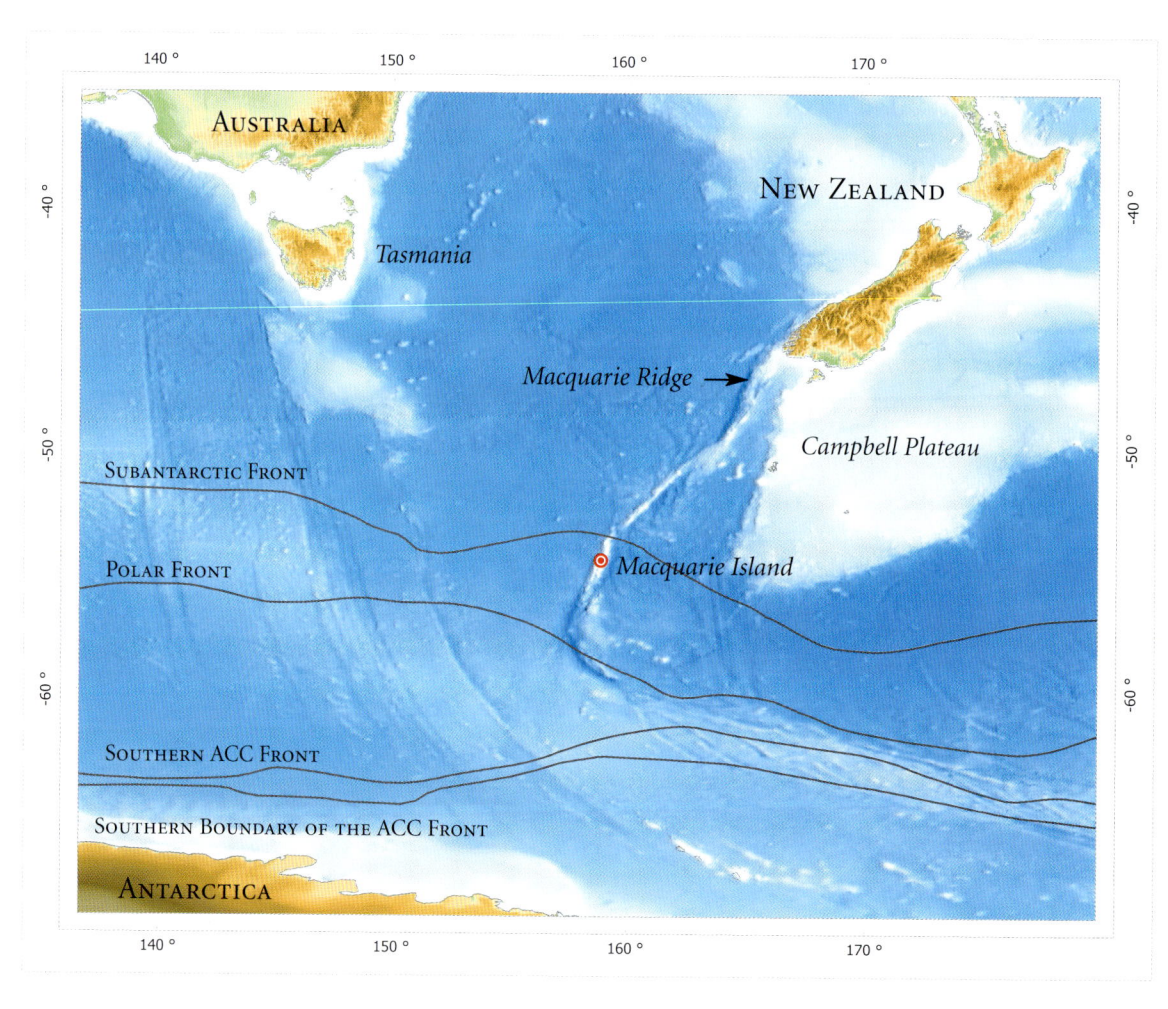

Such information is critical to the management of these species and monitoring the potential impacts of threats, such as fisheries.

The close relationship between how an animal feeds and its life history has long been recognised in a range of organisms. Albatrosses are no exception. They have evolved some extreme life history characteristics, such as delayed sexual maturity, slow reproductive rates and high rates of survival among adults. These characteristics have evolved as a response to a harsh and unpredictable lifestyle spent almost exclusively at sea, where storms are commonplace, temperatures often freezing, food scattered widely and distances unimaginably long. Before they can risk raising young, they must learn complex survival skills. Albatrosses use a range of strategies to find food and their breeding sites typically allow them easier access to that food. The fact that most of these islands on which albatrosses breed are in the subantarctic, in the vicinity of the great Antarctic Circumpolar Current and other highly productive regions, emphasises the important link between albatrosses and the distribution of food in the ocean.

ABOVE:
A Grey-headed Albatross checks out a potential mate on the southern slopes of Petrel Peak, Macquarie Island.

LEFT:
A female Wandering Albatross sits alert above her three-week-old chick just days before leaving it unattended for the first time.

MAIN ALBATROSS BREEDING AREAS

Species	TAS Offshore AUS	Macquarie Island AUS	Heard Island AUS	Iles Crozet/ Iles Kerguelen FRA	Marion Island RSA	Falkland Islands UK	South Georgia UK	New Zealand subantarctic islands	Chilean subantarctic islands
Wandering		•		•	•		•		
Shy	•								
Black-browed		•	•	•	•	•	•	•	•
Grey-headed		•		•	•		•	•	•
Light-mantled Sooty		•		•	•		•	•	

Source: Gales (1998), in Albatross Biology and Conservation. G. Robertson and R. Gales (eds). pp. 20-45. Surrey Beatty & Sons, Sydney.

A Royal Penguin stands on a Southern Elephant Seal at the edge of the Main Colony at Hurd Point, Macquarie Island. Although they generally co-exist quite happily, elephant seals can create chaos when they muscle their way through a packed penguin colony.

MACQUARIE ISLAND

THIS SMALL, ISOLATED SUBANTARCTIC island lies approximately 1500 kilometres south-east of Tasmania and 1100 kilometres south-west of New Zealand at 54°30'S and 158°55'E (see map page 20).

Only 5 kilometres at its widest point, it stretches approximately 34 kilometres from end to end running approximately north–south. It is bordered on both the western and eastern side by coastal slopes that rise steeply about 150–250 metres to the plateau edge.

The topography of the plateau is rugged with several prominent peaks rising to a height of 300 metres or more. Much of the land between the coast and the coastal slopes is dominated by wet, flat areas, particularly on the north-western side where it extends for several kilometres down the coast (right). In these places the water table is close to the surface and these boggy areas are referred to as the 'featherbed' for obvious reasons. Walking through them can be quite difficult and it is not uncommon to sink into it up to a metre or more. Such areas are also extremely fragile and susceptible to human damage so access to many of them is restricted.

Macquarie Island is an aerial exposure of uplifted ocean crust in the Southern Ocean and represents the highest point of the Macquarie Ridge complex, which extends to the north and, to a slightly lesser extent, the south of the island (see map page 20). The ridge marks the interface of two major tectonic plates and slopes down steeply on the western side of Macquarie Island to a depth of approximately 5000 metres (also known as the Macquarie Trench) approximately 20 kilometres offshore. Further to the west is a relatively featureless expanse that is comprised of the Emerald Basin, and further west again is the Campbell Plateau.[7]

Associated with this underwater topography is a dynamic region of fronts and zones that together comprise the Antarctic Circumpolar Current[8]. The northern boundary of the Antarctic Circumpolar Current is defined by the Subantarctic Front, which passes through a deep gap in the Macquarie Ridge to the north of Macquarie Island. This front represents the boundary between the warmer, salty waters of the Subantarctic Zone and the colder waters of the Polar Frontal Zone.[9,10] Further south the Polar Frontal Zone is bordered by the Polar Front, and close to the Antarctic shelf, two additional circumpolar fronts have been identified.[11]

Macquarie Island, with location of the main ANARE station, field huts and other sites of interest. Data provided courtesy of the Australian Antarctic Data Centre and NASA.

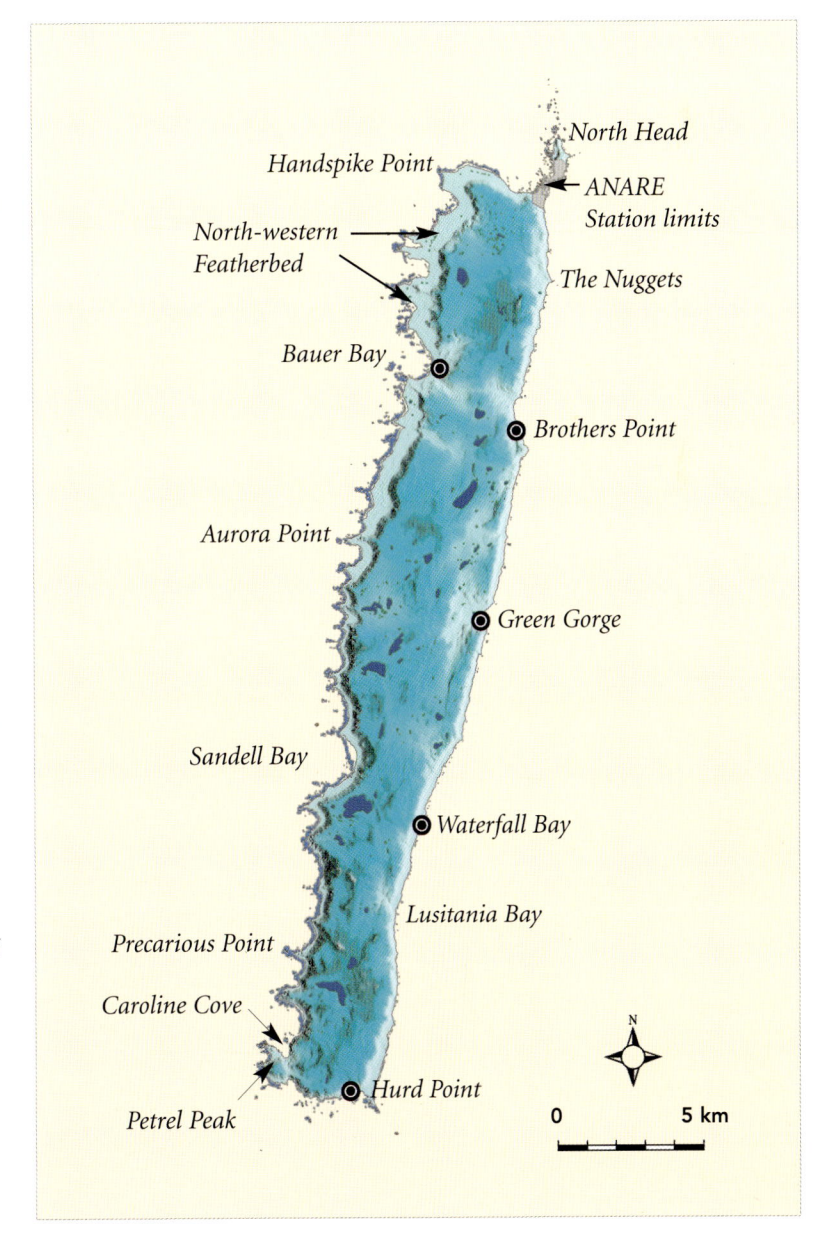

BREEDING AREAS 23

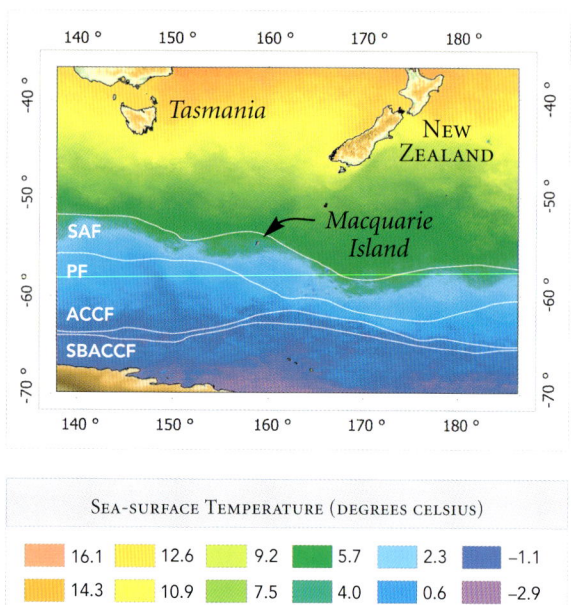

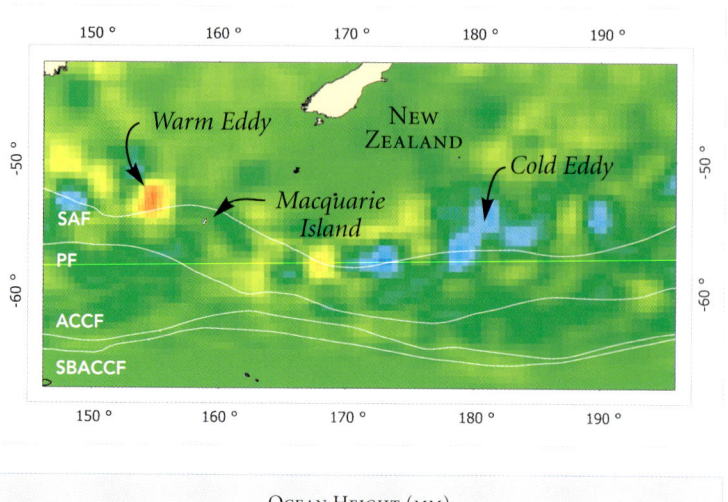

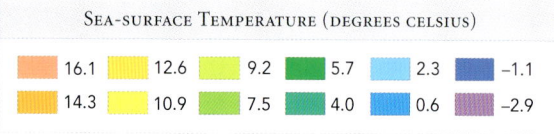

ABOVE LEFT:
Sea-surface temperatures around Macquarie Island and fronts of the Antarctic Circumpolar Current.[11,26]

ABOVE RIGHT:
Localised warm and cold eddies in the vicinity of Macquarie Island and fronts of the Antarctic Circumpolar Current.[11,27]

These define the southern boundary of the Antarctic Circumpolar Current; they are also shown on pages 18 and 20.

The different water masses that form these fronts and zones are often at different temperatures (above left), and within them there are also eddies and currents on a smaller scale (above right). The large global fronts, water masses of different temperatures and smaller-scale eddies and currents are fundamentally linked to the distribution of marine organisms, particularly plankton,[12,13] which are the building blocks upon which all oceanic food chains are based. There is often a strong link between what albatrosses eat and oceanographic features, and their food—mainly squid, small fish and/or krill—is not always easily found close to home. For example, while Black-browed Albatrosses breeding on Macquarie Island can find most of their required prey close to the island, some individuals still travel thousands of kilometres in a single journey to find food generally associated with shallower continental shelf waters. Others, like the Grey-headed and Light-mantled Sooty Albatrosses, regularly travel vast distances to the large fronts or smaller scale eddies where they are more likely to find their preferred food.

FLORA AND FAUNA

Wandering, Black-browed and Grey-headed Albatrosses breeding on Macquarie Island are largely restricted to the south-western corner. The majority of the Wandering Albatrosses breed in the Caroline Cove area, with most breeding pairs nesting in the Caroline Cove Amphitheatre or on Petrel Peak, and three to four pairs breeding on the north-western featherbed (see map page 23).

An unusually still day on the plateau. Pyramid Lake with a dusting of snow, Macquarie Island.

On the north-west coast of Macquarie Island the fragile environment of the wet coastal flats known as the 'featherbed' is bathed in the pinkish hues of sunset.

Black-browed Albatrosses only breed in the south-western corner, although there is a colony of over 100 birds breeding on the relatively inaccessible Bishop and Clerk islets some 30 kilometres to the south of Macquarie Island.[14] Black-browed Albatrosses also used to breed in the north of the island on the slopes of North Head but this small colony declined in the 1960s and 1970s and the last pair was recorded breeding in this area in the early 1980s.[15]

Grey-headed Albatrosses are restricted solely to the south-western corner of Macquarie Island, with most pairs breeding on the steep southern slopes of Petrel Peak, and just a few pairs breeding on the western slopes of Petrel Peak. Light-mantled Sooty Albatrosses are the most common breeding albatrosses on Macquarie Island and breed all around the island on the coastal slopes.

Albatrosses make up only a small part of the diverse range of animals and plants on Macquarie Island. Penguins are extremely abundant, particularly during the spring and summer months, and there are four species breeding there: Rockhopper Penguins (*Eudyptes chrysocome filholi*), Royal Penguins (*E. schlegeli*), King Penguins (*Aptenodytes patagonicus*) and Gentoo Penguins (*Pygoscelis papua papua*). The Royal Penguin is the most common penguin. During the summer months there are maybe over a million breeding pairs passing through the island. King Penguins are also common.

Their numbers are increasing following the decimation of the population by early oil merchants to the island in the 1800s.[7] Rockhoppers are the smallest penguin, but probably the most tenacious, and their attitude is reflected by numerous fights that often break out between individuals. Gentoo Penguins are the least common penguin on Macquarie Island. They breed not only on the beaches, but also on the grassy areas on the featherbed and even up onto the coastal slopes.

Albatrosses are the not the only members of the petrel family living on Macquarie Island. Several other petrels also breed there. Largest of these are the Northern Giant Petrel (*Macronectes halli*) and the Southern Giant Petrel (*M. giganteus*), which are similar in size to the smaller albatrosses. Southern Giant Petrels breed in colonies of just a few pairs to many

These Royal Penguins are gathered at the edge of a massive breeding colony that is home to nearly a million penguins in the summer months at Hurd Point, Macquarie Island.

BREEDING AREAS   25

RIGHT: Windswept tussock grass at Hasselborough Bay with snow-capped Mt Eitel in the background, Macquarie Island.

ABOVE: A dense field of the megaherb *Pleurophyllum hookeri* covers Bauer Creek, Macquarie Island.

CENTRE: A Gentoo Penguin stands to attention on the west coast of Macquarie Island.

RIGHT: This adult male Southern Elephant Seal lounges by Sandell Bay, Macquarie Island, awaiting the return of his harem of hundreds of females.

ABOVE: Breeding colonies of native Blue-eyed Shags are part of the landscape at Handspike, Macquarie Island.

LEFT: An emerging frond of the Macquarie Island fern, *Polystichum vestitum*, which sometimes provides a beautiful leafy habitat for Light-mantled Sooty Albatrosses.

LEFT: The south-west coast of Macquarie Island with the steep slopes of Petrel Peak in the background.

BREEDING AREAS

No peace for a small band of Royal Penguins at Bauer Bay, Macquarie Island, as a Subantarctic Skua searches for prey among weaker penguins.

hundreds of individuals and it is estimated that between 2000 and 2500 pairs breed each year on the island. Most colonies are around the coastal margins but there are some on other regions of the plateau. Northern Giant Petrels breed right around the coast of Macquarie Island and, in contrast to their close relative, they do not form major colonies: rather they breed either singly or in loose aggregations. There are approximately 1500 pairs of Northern Giant Petrels breeding on Macquarie Island each year.

Most of the other petrels that breed on Macquarie Island make their nests in burrows in the soil. These burrows range in size from the 1–2-metre-deep chambers of the Sooty Shearwater (*Puffinus griseus*) to the smaller burrows of the tiny Blue Petrels (*Halobeana caerulea*). Other species of burrowing petrel that breed on Macquarie Island include the Antarctic Prion (*Pachyptila desolata*), White-headed Petrel (*Pterodroma lessonii*) and Grey Petrel (*Procellaria cinerea*). The latter species was only recorded breeding on Macquarie Island in 2001 following eradication of feral cats, which had previously hunted them. Before this, no breeding pairs had been recorded for over 100 years and numbers increased steadily in the first few years following the first observed breeding attempt. Unfortunately habitat destruction caused by rabbits is now a major threat to these and other burrow-nesting petrels on Macquarie Island.

Other seabirds that breed on the island include the Subantarctic Skua (*Catharacta lonnbergi*), the endemic Blue-eyed Shag (or Cormorant) (*Phalacrocorax albiventer purpurascens*), the Kelp Gull (*Larus dominicanus*) and the Antarctic Tern (*Sterna vittata bethunei*). The Subantarctic Skua is a predatory gull-like bird and one of the few animals that can kill and eat albatross chicks when they are young. Their numbers are probably unnaturally high on Macquarie Island because they also feed on the rabbits, which are numerous. The Kelp Gulls and Blue-eyed Shags breed around the coast, the gulls feeding largely on small intertidal molluscs and invertebrates and the shags feeding in close inshore waters on small fish. Antarctic Terns are small, delicate-looking seabirds that nest around the coast and coastal sea stacks of Macquarie Island in small colonies. Several species of ducks have found their way to Macquarie Island and Common Starlings (*Sturnus vulgaris*) and Redpolls (*Carduleis flammea*) also breed there. Sightings of vagrant birds that have lost their way or been blown off course are relatively common.

Four species of seal commonly occur on Macquarie Island: three species of fur seal and the Southern Elephant Seal (*Mirounga leonina*).

Of the fur seals, only the Antarctic Fur Seal (*Arctocephalus gazella*) and the Subantarctic Fur Seal (*Arctocephalus tropicalis*) were thought to breed on the island but the New Zealand Fur Seal (*Arctocephalus forsteri*) 'hauls out' around the coast for a few months each year. During the summer months sub-adult and adult males of this species come ashore in large numbers and at such times the New Zealand Fur Seals become the most common fur seals on the island. There is also a distinct possibility that New Zealand Fur Seals are hybridising with the other two species of fur seals on the island. The original fur seal population was wiped out by sealers in the 1800s and they only started breeding again in the 1950s.[16] Since that time, recovery of the populations has been slow. Fewer than 200 pups are born each year,[17] although exactly how many pups are produced by each species is unclear due to possible hybridisation.

The Southern Elephant Seal was also hunted for its oil on Macquarie Island by sealers in the 1800s and early 1900s and today the population numbers some 60 000 individuals.[18] Southern Elephant Seals are polygamous, with adult male bulls—also known as beachmasters—forming and holding harems ranging in size from just a few individuals to hundreds of females. These harems occur right around the coast of Macquarie Island, with many of the biggest harems located on the isthmus close to where the main ANARE station is situated.

New Zealand Sea Lions (*Phocarctos hookeri*) are also occasionally observed around the coast of Macquarie Island; however, these are mostly males, with few females sighted. These sea lions have been seen hunting young fur seal pups as they venture into the water in the early weeks of their life.

Over 80 species of moss, more than 100 species of lichen and over 40 species of vascular plant occur on Macquarie Island. Three of the vascular plant species are endemic and five are thought to be introduced.[19] There are no trees or shrubs present due to the harsh climate. The island has a biogeographic affinity with Australasia, and similarities in vegetation are observed between Macquarie Island and some New Zealand sub-antarctic islands.

The three largest species—Tussock Grass (*Poa foliosa*), Macquarie Island Cabbage (*Stilbocarpa polaris*) and the megaherb from the daisy family (*Pleurophyllum hookeri*), which dominate the

LEFT:
The flowers of the Macquarie Island Cabbage (*Stilbocarpa polaris*) are usually yellow, but here are juxtaposed with a rare example of the red-flowering variety.

BELOW:
The vibrant green of a Coastal Cushion Plant (*Colobanthus muscoides*) highlights the otherwise harsh terrain of Hurd Point, Macquarie Island.

RIGHT:
Tangled kelp *Durvillaea antarctica* wraps its black, stringy arms over the intertidal zone, on the east coast of Macquarie Island.

BELOW:
Royal Penguins coming and going from the southern edge of the colony transform the bland rocky shores at Hurd Point, Macquarie Island, into a vibrant sea of black and white.

LEFT:
Protected by a rock stack on Eagle Point, Macquarie Island, a Northern Giant Petrel chick begs for food from its parent.

BELOW:
Macquarie Island Cabbage and tall tussock grass softens the otherwise rocky texture of the steep coastal slopes of Precarious Bay, Macquarie Island.

ABOVE:
A Northern Giant Petrel prepares to take flight from Caroline Cove Amphitheatre, Macquarie Island.

RIGHT:
A bare, snow-covered plateau during the harsh Macquarie Island winter.

RIGHT:
The dramatically beautiful Caroline Cove and Petrel Peak are home to most of the albatrosses on Macquarie Island.

BELOW:
Four species of penguins breed on Macquarie Island. Here, stately King Penguins (right) emerge from the water with a friendly Royal.

ABOVE:
A bloodied Southern Giant Petrel sits among the confetti of moulted penguin feathers after feeding on a seal carcass, Hurd Point, Macquarie Island.

LEFT:
Moulting Rockhopper Penguins at Caroline Point, Macquarie Island, leave a mess of feathers.

coastal slopes of the island—are also found on New Zealand subantarctic islands. The Cushion Plant (*Azorella macquariensis*), which dominates the windswept plateau at the higher altitudes, is considered an endemic species. There are three species of ferns. The most common, *Polystichum vestitum*, occurs on low-altitude, east-coast slopes, often in the vicinity of penguin colonies. There is also an orchid (*Nematoceras dienema*) which grows nowhere else and is the world's most southern orchid.[19]

Approximately a quarter of the island's plants are grasses, all from the Poaceae family. These vary in stature and habit. The Tussock Grass stands up to 1.8 metres high and dominates the tall grasslands that cover most of the island's coastal slopes. Short grassland communities are dominated by several species of grasses and small flowering plants. Raised coastal terraces—'the featherbed'—support waterlogged mire communities that are dominated by mosses, sedges and grasses. The visually striking and colourful herbfields are dominated by the Macquarie Island Cabbage and the daisy. These megaherbs are unlike the other plants as they are tall, growing to 80 centimetres high, with large leaves and flowers.

Rabbit grazing has had a negative impact on many plant species and it is likely that continued grazing since their introduction in the early 1800s has completely altered the original vegetation on the island. Although rabbit numbers were brought under control following the introduction of *Myxoma* virus in the 1970s, a suite of factors, including drier winters and reduced effectiveness of *Myxoma*, has resulted in a significant increase in rabbit numbers in recent years. This increase has resulted in large-scale vegetation damage and a significant reduction in much of the tussock grasslands on the coastal slopes. It has also severely damaged the Macquarie Island Cabbage and daisy in many areas.

MANAGEMENT AND RESEARCH

MACQUARIE ISLAND IS A NATURE reserve managed by the Tasmanian Parks and Wildlife Service in conjunction with the Nature Conservation Branch of the DPIWE, Tasmania. It is also an International Biosphere Reserve and a World Heritage Area. The Australian Antarctic Division has maintained an ANARE station on Macquarie Island since 1949 and the station continues to function as a base for research and maintenance. During 2004 and 2005, research on the island focussed on geology, climate change, albatrosses, seals and feral animal eradication. There is also an important meteorological station on the island. Due to the high diversity of wildlife and fragile areas of vegetation, access to much of the island is restricted, especially during the spring and summer months when many of the animals, including the albatrosses, penguins and seals are breeding.

The world's southernmost orchid, *Nematoceras dienema*, is exclusive to Macquarie Island. Photo: Tore Pedersen.

Speckling a promotory of Albatross Island, hundreds of Shy Albatrosses incubate their eggs and form the island's Main Colony.

Albatross Island

Albatross Island is a small island that lies approximately 30 kilometres north of the northern tip of Tasmania and 60 kilometres south-east of King Island.

The island is situated on the western side of Bass Strait in relatively shallow water of around 40 metres in depth (see map page 19). At just over a kilometre in length and approximately 300 metres wide at its widest point, it covers an area of 18.5 hectares and rises to a maximum height of approximately 60 metres. Its elongate shape runs north to south, and it has a convoluted coastline, bordered by steep cliffs on the eastern side and more sloping terrain on the west, although steep areas are still in abundance.

The rugged coastline is largely a result of the Precambrian conglomerate boulders and gulches and the geology is unique to Tasmanian islands.[20,21] Different rates of erosion of these areas have formed a succession of small rugged bays and caves right around the coastline. At the north of the island there are three main cave structures extending for approximately 300 metres at around 20 metres above sea level. It is likely that the force of the sea formed these caves over many tens of thousands of years.[21]

The continental shelf runs parallel to the island approximately 100 kilometres to the west. The large-scale movement of water masses against the continental shelf causes mixing of deeper and shallower waters and creates areas that are high in nutrients. These upwellings around the margins of the continental shelf are likely to play an important part in the distribution of food resources for albatrosses in this region and consequently in the feeding ecology of the Shy Albatrosses breeding on Albatross Island. The continental shelf and proximity of large landmasses strongly influences the prevailing currents around Albatross Island and there is a distinct seasonal contrast in the surrounding water temperatures. During summer, warmer waters extend down and across from southern Victoria and also down the east coast of Australia as part of the East Australia Current. This warmer water retreats in autumn leaving considerably colder waters around Albatross Island during winter (see map page 36).

FLORA AND FAUNA

Shy Albatrosses breed in four separate colonies on Albatross Island, named by early researchers as North, South, Main and West. Most of the birds breed in the Main Colony, which is in the middle part of the island. The colony extends for approximately 250 metres from north to south. Off to the side of the Main Colony, on its central western edge, is an area known as 'the trap'. This relatively small, steep-sided area probably formed after the collapse of one of the cave complexes. Its name derives from the tendency of birds to accidentally fall in, usually from a misjudged landing, and the difficulty of escaping. The colonies in the north and south of the island are smaller; smallest of all is the one on the western side of the island, where relatively few birds breed on more isolated nests.

Most information on the flora and fauna of Albatross Island has been obtained by Nigel Brothers, Rosemary Gales and their co-workers over the last 30 years. This information is well summarised in *Offshore Islands of Tasmania*, by Nigel Brothers and co-authors,[20] and is reproduced on the following pages.

A silhouette of Albatross Island at sunset, home to around 5000 breeding pairs of Shy Albatrosses. The gap is the result of a collapsed cave complex.

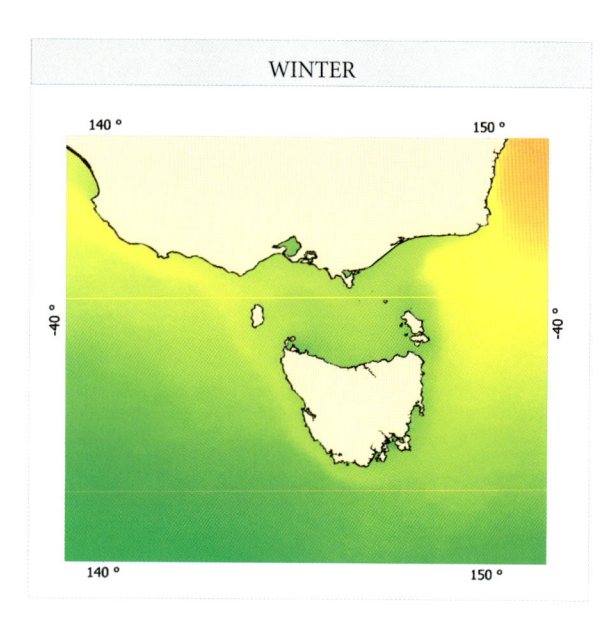

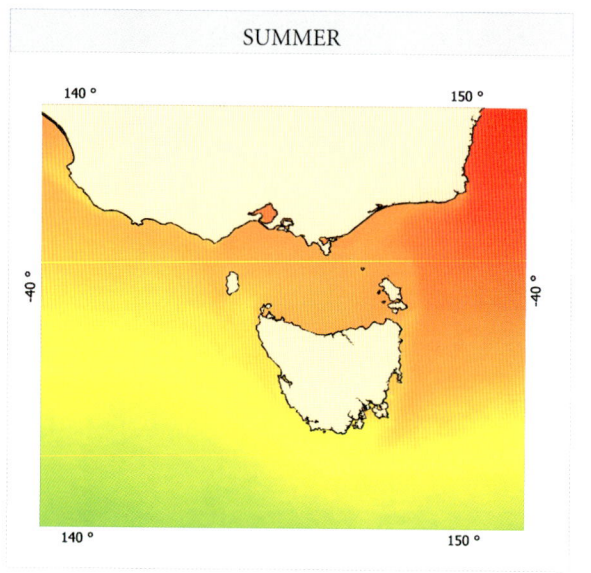

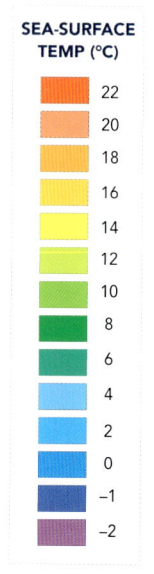

Winter and summer sea-surface temperatures around Tasmania,[26] showing the strong influence of the East Australia Current as it brings warmer water down the east coast of Australia.

The dominant seabird on Albatross Island is the Fairy Prion (*Pachyptila turtur*), with some 30 000–50 000 pairs estimated to nest there. This species utilises the extensive cave habitat to its fullest and occupies most nooks and crannies in these areas. Approximately 300 pairs of Little Penguins (*Eudyptula minor*) also nest in the larger cave areas and in small rock overhangs and ledges scattered around the island.

Other breeding birds include Short-tailed Shearwaters (*Puffinus tenuisrostris*), Pacific Gulls (*Larus pacificus*), Silver Gulls (*Larus novaehollandiae*), Sooty Oystercatchers (*Haemotopus fuliginosus*) and the White-bellied Sea-eagles (*Haliaeetus leucogaster*). Due to the island's proximity to the Tasmanian mainland, several other species of native bird are also regularly observed, but they do not breed there. These include the Brown Falcon (*Falco berigora*), Swamp Harrier (*Circus approximans*), Ruddy Turnstone (*Arenaria interpres*), Grey Fantail (*Rhipidura fuliginosa*), Forest Raven (*Corvus tasmanicus*) and Silvereye (*Zosterops lateralis*). Introduced birds that have been observed include the Skylark (*Alauda arvensis*), the Common Starling (*Sturnus vulgaris*) and the Common Blackbird (*Turdus merula*).[20]

Albatross Island was once a major haul-out and possible breeding site for fur seals before it was decimated by sealers in the 1800s.[22] Only occasional breeding occurs these days, but a small number of Australian Fur Seal (*Arctocephalus pusillus doriferus*) males regularly haul out on the eastern side. New Zealand Fur Seals also occasionally come ashore; however, there are no records of this species breeding here. The Metallic Skink (*Niveoscincus metallica*) is also present on Albatross Island in relative abundance, and the Tasmanian Tree Skink (*Niveoscincus pretiosa*) is also occasionally seen.

Albatross Island is relatively low in plant diversity. Most of the north of the island is covered by low-lying flowering herbs (*Disphyma crassifolium* and *Senecio sp.*). Other areas of the island are dominated by Blue Tussock Grass (*Poa poiformis*), and this vegetation type forms most of the Short-tailed Shearwater habitat. Shearwaters nest in burrows under the tussock, which provides insulation and some protection from would-be predators like sea eagles and ravens.

MANAGEMENT AND RESEARCH

Albatross Island is currently listed as a nature reserve and access to the island, whilst not prohibited at this stage, is discouraged due to the significant potential for wildlife disturbance. Because of the high density of albatrosses and the fragility of the ecosystem that supports burrowing petrels, it is vulnerable to the activities of people. An escaped campfire in 1982 killed approximately 900 Short-tailed Shearwater chicks, up to 100 Little Penguin chicks and many Fairy Prion chicks.[20] Its nature reserve status is absolutely necessary to protect this wild island as it slowly recovers from the historical impacts of humans.

ABOVE: The natural amphitheatre known as 'the trap', appropriately named for the tendency of birds to fall in and be unable to get out. The crowded Main Colony is in the background.

LEFT: Shy Albatrosses soar above the rocky shoreline of Albatross Island as they return to their nests at the end of the day.

BREEDING AREAS

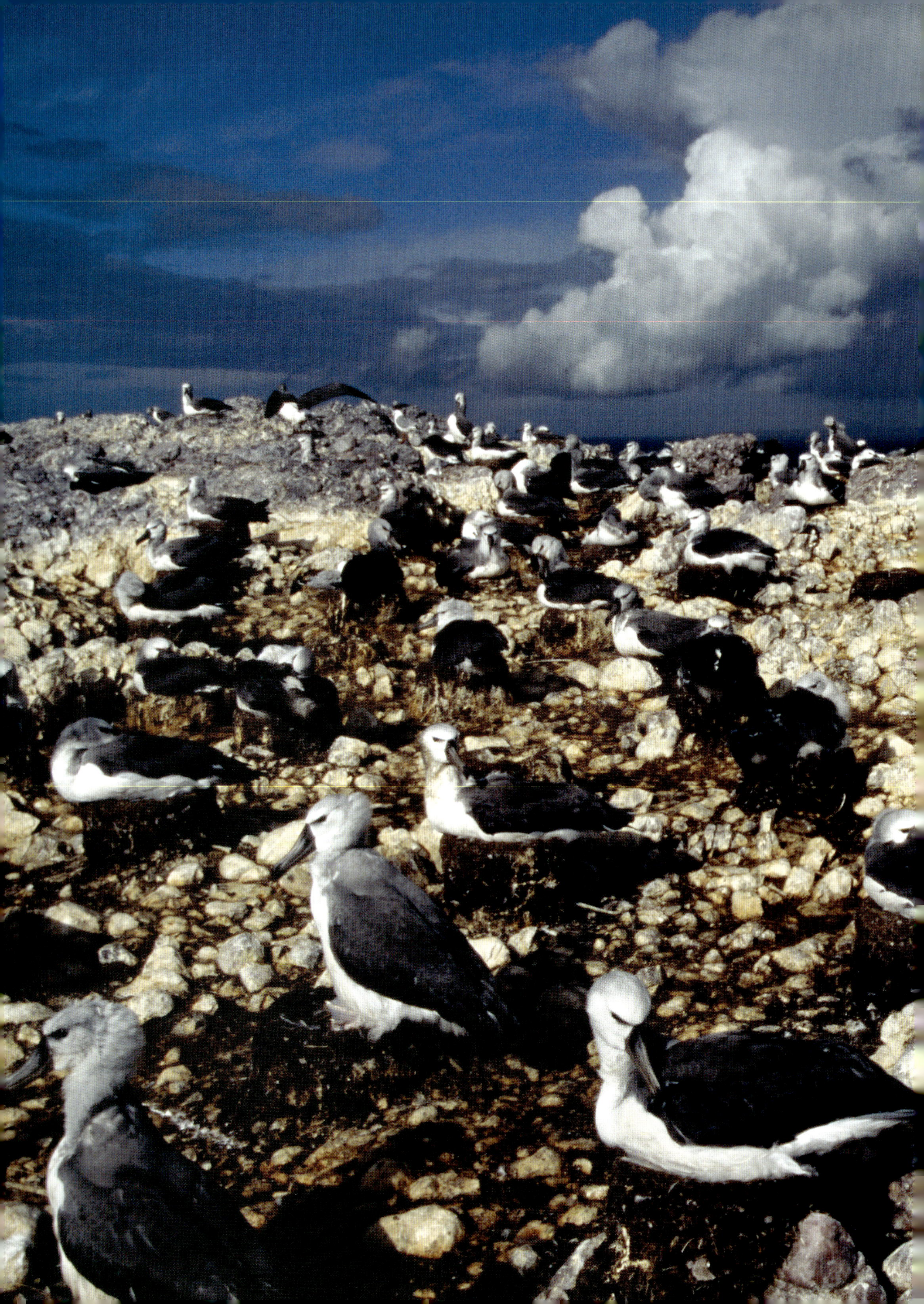

Shy Albatross colonies are very crowded, like this one in North Colony on Albatross Island. There is barely enough space between nests for the birds to stretch their large wings.

These Shy Albatross chicks will fledge in two to four weeks. They blend in with their surroundings on the rocky, guano-splattered slopes of Pedra Branca.

Pedra Branca

The tiny island of Pedra Branca lies 20 kilometres due south of the southernmost point of Tasmania. And, together with the Mewstone, it is much closer to the edge of the continental shelf than Albatross Island.

Less than 50 kilometres from its coastline, the seabed falls sharply away to water over 2000 metres deep. During the summer months the island is influenced by warm currents coming down the eastern and western side of Tasmania but during winter, like Albatross Island, these waters retreat, and cooler waters from the Southern Ocean dominate the area.

Pedra Branca is only 270 metres long by 100 metres wide and rises to a maximum height of approximately 55 metres. It is wedge-shaped, with steep terraced rocky slopes on both western and eastern sides. The central ridge, formed by the apex of the east and west sides, runs in a general north–south direction, so the west side bears the brunt of the almost ubiquitous south-westerly and westerly swells.

Pedra Branca is largely composed of sedimentary rock about 40 metres thick, overlain by a dolerite sheet.[23] There is also a unique geological feature known as 'the Rockpile', which is located on a western ledge about 20 metres above sea level. It is likely that this conglomeration formed sometime between 25 000 and 100 000 years ago.[22] After it formed, phosphatic and siliceous solutions (largely derived from guano) probably flowed into it through the adjacent sandstone wall,[23] creating a solid mass of cryptic holes and crevices with jagged edges that form a near perfect habitat for the Pedra Branca Skink (*Niveoscincus palfreymani*).

FLORA AND FAUNA

Of the three Tasmanian islands, Pedra Branca has the smallest breeding colony of Shy Albatrosses with approximately 200–250 pairs breeding there each year.[20] The nests are spaced out along the sloping side of the eastern ridgeline, with the two densest areas, in the middle and south of the island, known as the Main Colony and South Colony. Australasian Gannets (*Sula serrator*) also breed in large numbers on Pedra Branca and it is estimated that 6000–8000 pairs breed on the island each year.[20] Due to the lack of nesting material, many of the gannet nests are composed of rope, small

Pedra Branca, encrusted with generations of gannet guano, rises from the rough waters of the Southern Ocean.

BREEDING AREAS   41

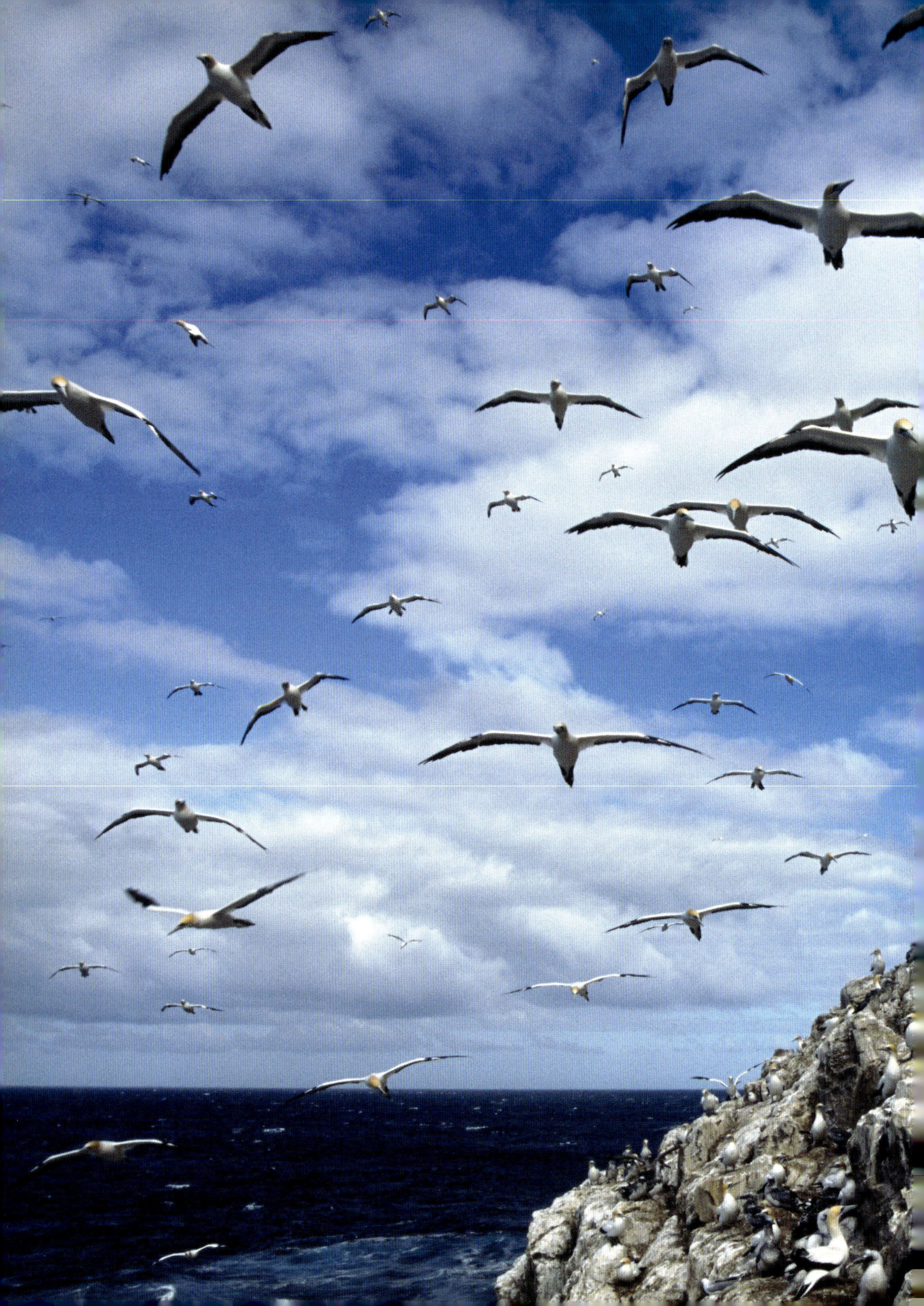

Gannets flying over their crowded breeding colony on the upper slopes of Pedra Branca. Soon their chicks will be ready to take flight and leave behind their rocky island home.

bits of netting and other marine debris. It is a strange sight to see such an abundance of human waste material on one of the least-visited islands in the Southern Ocean—and a sad testament to the amount of debris that must be floating in the water. Other breeding birds on Pedra Branca include the Pacific Gull, Silver Gull, Kelp Gull and the Black-Faced Cormorant (*Phalacrocorax fuscescens*).

Pedra Branca is probably best known for its endemic skink. Although superficially similar to some Tasmanian species, the Pedra Branca Skink occurs nowhere else in the world. It is relatively large, with adults attaining a length of up to 20 centimetres and weighing 15 grams. Most of the skink population lives in 'the Rockpile' on the main western ledge and, even though they are subject to regular soaking by the strong westerly and south-westerly swells, the skinks are able to survive by moving deep into the holes, cracks and crevices that this habitat provides. These skinks also live in other areas of the island, where they also prefer cryptic areas with deep cracks or crevices: these afford them protection from their main predator on Pedra Branca, the Silver Gulls.

Population estimates of the skink have varied over the last decade and the population monitoring is under review, but current estimates suggest that there may be as many as 400 individuals on the island. The skinks feed predominantly on small invertebrates, but also on the regurgitated food of gannets during the summer months.[24] Due to its small population, restricted distribution and evidence of a decline in the 1990s, the species is listed as *Vulnerable* under the *Environment Protection and Biodiversity Act 1999* (Commonwealth) and as *Endangered* under the Tasmanian *Threatened Species Act 1995* (Tasmania).

Pedra Branca is also a haul-out spot for Australian Fur Seals, which occasionally come ashore in their hundreds along the lower rock platforms during calmer sea conditions. These seals are dominated by adult and subadult males that tend to gather together in large numbers at this and other island haul-outs around the coast of Tasmania. Occasionally New Zealand Fur Seals also come ashore here.

There is very little soil on Pedra Branca due to its small size and the frequency of the wave wash. As a result vegetation is very sparse, and only one living plant, the Beaded Glasswort (*Salicornia quinqueflora*), exists. It occurs only in very small numbers and is restricted to cracks in the rock.[20]

MANAGEMENT AND RESEARCH

PEDRA BRANCA IS A national park and a permit is required to access it. It is also listed as a World Heritage Area and the wildlife present makes it high in biological value. Its remote location, the difficulty of access and the often extreme prevailing wind and swell conditions make it a little-visited island. The Nature Conservation Branch of the Tasmanian Department of Primary Industries, Water and Environment has been conducting research there since the early 1980s and April is typically the only time that the island is visited by researchers each year.

Gannet chicks look very different from their parents before the chicks acquire their sleek adult plumage, which may take one to two years.

A Pedra Branca Skink, exclusive to the island, soaks up the sun on the convoluted 'Rockpile'.

Shy Albatross chicks, still with a head of down, sit on their lonely Mewstone nests. Unlike other species, some Shy Albatrosses lay their eggs straight into depressions in the rocks, with minimal nesting material. In the background is Maatsuyker Island and the south coast of Tasmania.

The Mewstone

The Mewstone is an impressive pyramid-shaped rock rising out of the Southern Ocean approximately 20 kilometres south of Tasmania and 10 kilometres to the south-east of Maatsuyker Island. Like Pedra Branca, it is close to the edge of the continental shelf, and less than 50 kilometres from its coastline the seabed falls sharply away into water over 2000 metres deep. It is subject to the same strong seasonal differences in water temperature as Pedra Branca, with warmer currents coming down the eastern and western side of Tasmania during the summer months and retreating during winter.

At the south-eastern end of the Mewstone the small boulder-strewn summit rises to an impressive height of 140 metres. The island covers an area of just over 13 hectares, with a smaller islet of 1.7 hectares—sometimes known as 'the Mewlet'—adjacent to the southern end. A jagged ridge-line runs south-easterly up to the summit from the steep cliffs that mark the northerly point. A second, less-defined ridge-line runs south-south-west from the summit to cliffs at the southern end. The eastern and western slopes are terraced in most of the upper sections before dropping away into steep cliffs in the lower sections.

The Mewstone is composed largely of granite, with some quartz veins, and its rugged terrain is typical of its geological formation.[23] It is estimated that the Mewstone was isolated from the Tasmanian mainland as the sea level rose at the end of the last glacial period some 15 000 years ago and its geology is unique in that none of the surrounding islands are composed of this type of rock.[21]

A punishing south-westerly swell wraps around the side of the Mewstone, spotted with white specks that are thousands of nesting albatrosses.

Guano from nesting albatrosses and cormorants streaks the north-eastern side of the Mewstone.

ABOVE:
An adult Shy Albatross prepares to take flight from the Mewstone.

LEFT:
The small islet known as the Mewlet stands exposed to the east of the Mewstone.

ABOVE:
The black bill of this Shy Albatross chick will slowly turn yellow in its first few years of life.

RIGHT:
These Shy Albatross chicks on small mud nests in the convoluted rocks that form the main ridge of the Mewstone are almost ready to fledge.

BREEDING AREAS 49

FLORA AND FAUNA

ALBATROSSES DOMINATE THE WILDLIFE on the Mewstone and it is estimated that about 7000–7500 pairs nest there each year.[20] The Mewstone hosts the largest colony of Shy Albatrosses. This colony has not suffered from any human exploitation, unlike those on some of the more accessible islands.

Although albatrosses nest right around the island, most breed on the more terraced western side in dense aggregations. Most nests are in the upper two-thirds of the island, as most of the lower cliffs are too steep for albatross nesting. Many pairs also nest on the jagged ridge-lines where the convoluted terrain allows pairs to find more suitable ground relative to the steeper sections. Because there is so very little soil, many eggs are laid in depressions of the rock with little or no surrounding nest material.

Fairy Prions also nest in the numerous cracks and crevices right around the Mewstone and, although the terrain makes it difficult to accurately determine the number of breeding pairs, it was estimated in the 1980s that some 20 000 pairs were nesting there.[20] The cryptic nature of their nesting sites and the inaccessibility of much of the island make obtaining accurate counts of this species difficult. Several Silver Gulls have been observed nesting on the lower cliffs, particularly on the southern side. Black-faced Cormorants also nest in small numbers on the eastern side and on the small islet to the south.

The Tasmanian Tree Skink (*Niveoscincus pretiosa*) is present all over the island and is often seen, particularly on hot days. Occasionally fur seals—generally sub-adult male Australian Fur Seals—haul out on the lower, flatter sections of the Mewstone and the accompanying islet. Due to the sheer nature of the lower sections, seal numbers are generally low on the Mewstone itself.

Because of the lack of soil and the exposed nature of the island there is very little vegetation on the Mewstone. In the few spots where soil does accumulate, Blue Tussock Grass is one of the most abundant species. Small areas of the herb *Senecio lautus*, Pigface (*Carpobrotus rossii*), the fern *Asplenium obstusatum* and succulent *Salicornia quinqueflora* are also present.[20]

MANAGEMENT AND RESEARCH

THE MEWSTONE IS A national park managed by the Tasmanian Parks and Wildlife Service. As with Pedra Branca, a permit is required to access this island and it is rarely visited, in order to maintain its relatively pristine state. Researchers from the Nature Conservation Branch of the Tasmanian Department of Primary Industries, Water and Environment conduct day trips to the Mewstone once every year or two to maintain and count some of the study sites that have been established to assist in the remote monitoring of Shy Albatross numbers.

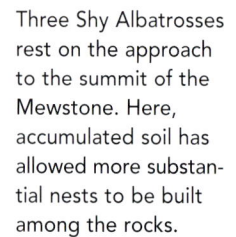

Three Shy Albatrosses rest on the approach to the summit of the Mewstone. Here, accumulated soil has allowed more substantial nests to be built among the rocks.

This Shy Albatross fledgling is on a nest near the summit of the Mewstone. Most nests are near the summit as the lower cliffs are too steep for albatross nesting.

ABOVE:
A drawing of elephant seals on King Island, 1767–1814 (Nouvelle-Hollande, Ile King, l'elephant-marin ou phoque a trompe (*Phoca proboscidea*, N), vue de la Baie des Elephants). Victor Pillement. National Library of Australia.

RIGHT:
A Subantarctic Fur Seal, most probably the main species on Macquarie Island before sealing, rests in the Tussock Grass on the edge of the beach, on the east coast of Macquarie Island.

Humans in Bass Strait and the Southern Ocean

The seventeenth and eighteenth centuries were a time of rapid global expansion as Europeans, mostly from Britain, France and Holland, explored the oceans of the world, looking for land to claim as their own.

On the voyages of these early explorers the waters of the southern hemisphere were penetrated and mapped. The seafarers returned to their mother countries with tales of the sights and potential resources that abounded in these southern seas. The voyage of the *Endeavour*, captained by James Cook, was one such exploratory journey. Between 1768 and 1771 Cook made his way to Tahiti, New Zealand and Australia. On his return to England in 1771, it was Tahiti that really sparked the public imagination, with New Zealand and Australia initially thought to be of little interest.[1] Cook was also the first to fully explore the great Southern Ocean, when in 1772 he ventured upon an epic navigation that took him further south than any human had ever been before.[1]

Eight years after Cook's landing at Botany Bay, the British government began to contemplate the settlement of Australia as a penal colony. The decision to colonise Australia was also influenced by economics. The first settlers were from a Georgian England in the throes of the Industrial Revolution: natural resources were seen as being there for the taking and conservation as we know it today was not a common concept. When these first settlers reached Tasmanian waters and eventually Macquarie Island in the Southern Ocean, their impact was widespread and dramatic.

Industries such as whaling and sealing had already been well developed in the northern hemisphere, and it was partly the competition for these resources in more northern waters that provided the impetus for expansion into southern latitudes. These industries had devastating impacts on marine mammal populations in Tasmanian waters and on Macquarie Island. In addition, the hunt for oil led to the demise of many thousands of penguins on Macquarie Island, and feather hunters on Albatross Island reduced the Shy Albatross population to near-extinction levels.

Although the exploitation of marine mammals and seabirds has largely ceased in Tasmanian waters and in most areas of the Southern Ocean, humans are still exploiting the ocean with widespread fishing practices that are seriously harming many seabird populations around the world. Since large-scale long-line fishing began in the late 1950s and early 1960s, thousands upon thousands of albatrosses and smaller petrels have been caught and killed on fishing hooks, mostly from vessels fishing for tuna and, in more recent times, Patagonian Toothfish.

One of the most effective ways to catch both tuna and Patagonian Toothfish is to use a series

The rusty remains of Penguin boilers at The Nuggets, Macquarie Island, where thousands of penguins met their death. Photo: Justine Shaw.

A long-line fishing vessel attracts hundreds of albatrosses and petrels as it prepares to set its hooks. Photo: Graham Robertson.

of long lines with baited hooks. Long-lining involves setting baited hooks on lines that may be up to 100 kilometres long. The danger to seabirds comes from the setting of the hooks when they are first released from the vessel. The hooks do not sink below the surface immediately, and the bait on the hooks presents itself as an irresistible source of food to albatrosses and petrels. When the birds attack and swallow the baited hooks, they are hooked and then drowned as the weight of the line continues to take the hook down into deeper waters. Any non-targeted animals that are accidentally caught or killed while fishing—such as these birds—are known as by-catch.

Such widespread mortality has not been restricted to the areas around which the albatross breed and, given the global scale over which many albatrosses forage, in conjunction with the extensive nature of the fisheries, the by-catch of albatrosses has occurred in most oceans of the world. Conservation efforts have helped to reduce the level of seabird by-catch in many parts of the world, and many fishers now use measures to reduce the likelihood of catching and killing seabirds. These methods include the use of streamer lines to scare the birds off, setting gear at night when most albatrosses do not feed, using weighted lines to make the baited hooks sink more quickly and reducing the offal discharge that attracts birds to the vessels. Nevertheless, many populations around the world continue to decline and ongoing conservation efforts are needed to manage fisheries and reduce the level of seabird by-catch.

WHALING

Humans have killed whales for food since prehistoric times, when they used naturally stranded or trapped whales in coastal waters for food and blubber. Blubber is the whale's insulation and contains high levels of fat that can be used as both food and fuel. Several indigenous cultures are well known for their whaling practices. The Inuit and Indians of North America and the Ainu people of Japan have traditionally carried out seasonal whaling for centuries[2] but it was small scale and probably sustainable.

Large-scale commercial whaling began in the northern hemisphere. The Basque people of the Spanish coast are credited with starting whaling on a large scale as early as the ninth century—but it wasn't until the 1400s that they started pioneering long-distance whaling voyages.[2] Over the next 200 years whaling practices quickly expanded to include the waters around Iceland and Greenland and down into the Atlantic Ocean off Newfoundland and Canada.

By the 1600s the large whale stocks in the Arctic had been discovered and soon British ships were competing with the Dutch to exploit as much of this fertile new resource as possible. Whaling fleets from North America were formed soon after this area was colonised by the British, and through the first part of the eighteenth century these vessels began hunting in the southern hemisphere, taking whales off the coasts of Brazil and Argentina. Following the American Civil War, British ships also began to expand into the South Atlantic. Most of the interest in whales at this time was related to the oil, which could be extracted from not only the blubber but many other parts of the whale's body, including bones, tissue and other organs.

Expansion of whaling into the South Pacific was slower as the British East India Company had the sole rights to trade east of the Cape of Good Hope and guarded its monopoly jealously.[2] The settlement of Australia was not only seen as an ideal solution to dealing with the increasing number of criminals in England; it was also a good way to protect these eastern trading routes from the French.[1] Exploratory Dutch vessels had also reached the large southern continent and, by the late 1700s, intrepid American whalers from the other side of the Pacific had penetrated the waters of the Derwent Estuary in Tasmania.[1] The discovery of exploitable wildlife populations in Bass Strait, and concerns about the French and Americans in Tasmanian waters, led to the Governor of New South Wales at the time, Philip King, deciding to establish a colony on the River Derwent in Tasmania.[1] Whaling quickly became economically important to the new settlement, and one of the reasons given for the movement of the first settlement from Risdon Cove to Sullivans Cove in the River Derwent was the commercial advantages of the new harbour for whaling vessels.[2]

Anecdotal reports from this time talk of whales in the River Derwent being so abundant that small boats were forced to keep close to the shore to avoid being upset by the passage of whales.[3] Indeed, there were so many whales in

A blue whale and whaling men on the flensing or cutting-up plan at the Grytviken whaling station, Prince Edward Cove, South Georgia, Shackleton expedition, 1914–17. Photo: Frank Hurley. National Library of Australia.

ABOVE
Whaling, Lady's Bay, Tasmania c. 1848. Oil on canvas. William Duke. The Queen Victoria Museum and Art Gallery.

RIGHT
Flinching a yearling, a young sea elephant, Tristan De Acunha, 1824. Watercolour. Augustus Earle. National Library of Australia.

this river that the sound of the whale's blows was apparently enough to keep people awake at night. These whales were Southern Right Whales (*Eubalaena australis*), so named by the first Europeans as they were the 'right' whale to catch. During the first four decades of the 1800s this species formed the primary catch of the Tasmanian industry.

Southern Right Whales are relatively slow-moving animals that could be approached and killed by men in open boats. Also, usefully, these whales were positively buoyant and so floated after being killed, allowing for much easier handling. They also yielded large quantities of good-quality oil. Sperm Whales (*Physeter macrocephalus*) were also considered a valuable source of oil but they do not often venture as close to the shore, so they could not be caught so easily from shore-based whaling stations of the type that were set up on the River Derwent in the early 1800s.

Southern Right Whales seen in Australian waters undertake an annual migration. They move northwards from their Southern Ocean feeding grounds after the summer to breed in warmer waters around the Australian mainland during the winter. In the 1800s, when whaling was in full force, these whales used to breed in Tasmanian waters, and it was these breeding aggregations that made it easy for them to be caught and killed by the early whalers. At least 20 shore-based whaling stations were set up around the south and east coasts of Tasmania[2] to take advantage of these aggregations and during the first three decades following settlement, the export of whale oil was one of the biggest sources of wealth for the new Tasmanian economy.

It has been estimated that some 25 000 Southern Right Whales were killed in Australasian waters during the nineteenth century, with almost half caught between 1839 and 1895 and just under a third caught in Tasmanian waters.[4] Today it is estimated that only 1000 Southern Right Whales move through Tasmanian waters each year.

Whaling ceased to be a major industry in Tasmanian waters around the end of the nineteenth century as numbers dwindled to the point where it was no longer economically viable to catch them. In 1935 the precursor to the United Nations, the League of Nations, outlawed the hunting of Southern Right Whales and, apart from some illegal hunting by foreign vessels, whaling is thought to have stopped in Australian waters around this time.[2]

Although very few Southern Right Whales have been recorded breeding in Tasmanian waters since the decimation of the population, sightings of this species over recent years in the River Derwent have increased. Nevertheless, numbers are still incredibly low compared to what they once were, and the impacts of these early whalers are still clearly felt today.

SEALING—TASMANIA

IN ADDITION TO THE WHALING industry, the early European settlers also hunted seals in great numbers. Before the arrival of these settlers, four species of seal bred in Tasmanian waters: the Australian Fur Seal, the New Zealand Fur Seal, the Southern Elephant Seal and the Australian Sea Lion (*Neophoca cinerea*).

The *Emily Downing*, 1879, a whaling ship, docked at the new wharf at Hobart Town unloads whale oil casks. State Library of Victoria.

A moulting female Southern Elephant Seal at Caroline Cove, Macquarie Island. These seals were hunted on a large scale for their oil during the nineteenth century.

Whilst it is likely that early indigenous people living in coastal areas of Tasmania used seals for food and other resources, the large-scale slaughter really got underway in the late 1700s with the arrival of the first Europeans. This large-scale sealing began in Australian waters in 1798 when Captain Charles Bishop in the vessel *Nautilus* started hunting in Bass Strait. In just a few weeks of hunting, over 9000 skins were collected[5]—this heralded the beginning of a new era in the exploitation of animals in Bass Strait waters.

Seals in Bass Strait were exploited for two reasons. The elephant seals were killed for the oil that could be obtained by rendering or boiling down the carcass, and the thick, hairy coats of two species of fur seal and the sea lion could be exported for use in the clothing industry. Today, there are nine primary breeding sites of Australian Fur Seals in Bass Strait—prior to the arrival of the sealers, there were probably many more. Islands such as Albatross Island, where Shy Albatrosses breed, once supported dense aggregations of seals but today they are no more. When the sealers undertook the mass cullings, they stayed on the breeding colonies for weeks or even months at a time. On places like Albatross Island their presence would have impacted on the other wildlife living there, with birds like albatrosses and shearwaters being eaten and albatrosses being killed for their feathers.[5]

The number of seals then populating Bass Strait is hard to imagine given their relatively sparse numbers today. With large breeding colonies of fur seals on many Bass Strait islands and dense aggregations of Southern Elephant Seals on King Island, the sealing industry proved extremely lucrative. Over 200 000 animals are thought to have been taken from Bass Strait and King Island through the 1800s.[6] Despite some of the early Europeans foreseeing the extinction of the species under such intense hunting pressure, economic greed and competition between the sealing gangs meant that little restraint was employed and sealing continued until there were not enough seals left to make the industry commercially viable.

In 1808, barely a decade after sealing began, many of the larger sealing merchants stopped sealing due to a drop in skin prices and a reduction in the numbers of seals. This also coincided with the discovery of untapped sealing grounds off the coast of New Zealand and on subantarctic islands (including Macquarie Island). Smaller operators continued to take seals up until the 1830s, but by this time numbers had become so diminished in Bass Strait that seal hunting was not viable for even these 'lone sealers'.[5]

At the cessation of commercial sealing, the Southern Elephant Seal, Australian Sea Lion and New Zealand Fur Seal had been completely eliminated from Bass Strait and only very small numbers of the Australian Fur Seal remained. Elephant seals and sea lions have never returned to breed in Bass Strait and, while numbers of the Australian Fur Seal have slowly increased, they are still far below the estimates of pre-sealing numbers.

SEALING—MACQUARIE ISLAND

SEALING ON MACQUARIE ISLAND began as ships began to venture further afield, exploring uncharted waters in the Southern Ocean, following the demise of the large-scale Tasmanian industry. In 1809, the vessel *Perseverance* left Sydney on an exploratory voyage to look for exploitable stocks of seals in the Southern Ocean. The *Perseverance* carried one of the most experienced sealing gangs to ever leave Sydney. They briefly visited the Auckland and Campbell Island groups in New Zealand's subantarctic before reaching the shores of Macquarie Island on 11 July 1810.[7] Eight men and sufficient provisions for nine months were landed. On seeing the multitude of wildlife, including elephant seals and fur seals, the *Perseverance* immediately set sail for Sydney for fresh supplies and men.

So began the sealing industry on Macquarie Island and the large-scale destruction of the seals. Due to the effectiveness and the ruthlessness of the early fur seal hunters it is unlikely that we will ever really know the species composition or number of the original fur seal populations on Macquarie Island. In a similar manner to the Tasmanian industry, the exploitation was quick and comprehensive. Fur seals were eliminated approximately 10–15 years after their discovery in 1810[7] and they were not recorded breeding again on Macquarie Island until the mid-1950s.[8] Cargo records indicated that just under 200 000 skins were taken from Macquarie Island.[6] If that number of animals were present, annual pup production may have numbered around 40 000–50 000 individuals. While the identity of the main species of fur seals in the island prior to sealing is not known, it is likely that they were Subantarctic Fur Seals.[9]

The Southern Elephant Seal population on Macquarie Island was also vastly reduced during an extended period of exploitation from the 1800s to the early 1900s.[7] There was a steady

Sealers rolling barrels up the beach at The Nuggets, the main camping ground on Macquarie Island. The barrels were to be filled with penguin oil. State Library of New South Wales.

This male New Zealand Fur Seal lounging on Hurd Point, Macquarie Island, sports characterisically long whiskers.

ABOVE
Seal skins on deck of ship, 1907–31. State Library of Victoria.

RIGHT
A bull Southern Elephant Seal powders himself with sand at Caroline Cove, Macquarie Island. This habit is thought to assist in thermoregulation.

HUMANS IN BASS STRAIT AND THE SOUTHERN OCEAN

demand for animal oils for house and street lighting, and the oil from an elephant seal was considered of good quality, only just inferior to that of a Sperm Whale. The larger size of the elephant seal and the difficulties in transporting the blubber to the 'trypots' where it could be boiled down slowed the slaughter of this species so, unlike the fur seals, this industry continued for many years. The difference in size between the males and females also helped slow the decline of this species. The larger males would have undoubtedly been targeted since they produced significantly more oil than the smaller females. The men used to kill the sleeping animals, cut off the blubber with a knife and put it in a boiler on rocks and make a fire underneath with lumps of the same blubber.

The oil was then poured into barrels and eventually shipped off the island and sold in New South Wales or Britain. Thousands of tons of oil from elephant seals were removed from Macquarie Island during the 1800s.[7]

Conditions were extremely rough for the sealers living and working on Macquarie Island. For many months at a time they lived in exceptionally squalid conditions. They typically made huts lined on the inside with the furs of the seals and on the outside with Tussock Grass or the hides of elephant seals.[7] The men also used the seal furs as clothing to protect themselves from the often harsh Macquarie Island climate. Their only source of light and heat was burning the oil and blubber from the seals and the inside of their dwellings was black

Relics of the barberous seal hunters lie scattered at an outdoor museum on Macquarie Island. Photo: Felicity Jenkins. National Library of Australia.

with the smoke.[7] They had the advantage of at least some local edible vegetation—the Macquarie Island Cabbage probably saved many of them from scurvy.

Several different sealing gangs visited Macquarie Island throughout the 1800s. By 1890 elephant seals were becoming scarcer around the more accessible parts of the island and the removal of oil was not proving to be as commercially viable.[7]

The sealers had a significant impact on Macquarie Island. Not only did they eliminate most of the breeding seals; they also fed on the other animals living there. In addition, they brought with them introduced animals. Some of these remain on the island today.

SEABIRDS

EVEN SEABIRDS DIDN'T ESCAPE the attention of hunters around Tasmania and on Macquarie Island. Not only were the eggs of the Shy Albatrosses breeding on Albatrosses Island used for food but sealers eventually realised that the feathers of the albatross could also be sold. By the mid-1840s thousands of birds were probably being taken each year.[10] The sealers on Albatross Island used to catch the albatrosses by knocking them down with sticks. To make the job easier they drove them down into 'the trap', a relatively large natural amphitheatre with steep sides that the albatrosses had difficulty getting out of. Once in the trap the albatrosses were at the mercy of the hunters, who then stunned and plucked them, often leaving them to linger on to a slow death.[11]

The combined effects of sealers eating their eggs and the activity of feather hunters must have had a devastating impact. Ashworth and le Soeuf, on a visit to Albatross Island in 1895, described there being 400 nests scattered about in various places around the cliffs.[12] This differs greatly from the reports of George Bass and Matthew Flinders, who were probably the first Europeans to visit the island, in 1798, and described the top of the island as being 'almost white with birds … covered in albatrosses sitting closely on their nests', so closely, indeed, that George Bass was forced to form a passage through them with his seal club.[13]

Muttonbirds were also killed by the early sealers. Both the eggs and the chicks were eaten, and together they formed a major component of the sealers' food. Short-tailed Shearwaters were taken on the Bass Strait Island and Sooty Shearwaters were taken on Macquarie Island. As seal numbers diminished in Bass Strait, muttonbirding became an increasingly important industry for those sealers who remained. In addition to their use as food, their oil was used for lamps and their feathers became a useful and saleable commodity. It is likely that hundreds of thousands of shearwaters were taken from the islands around Flinders Island (north-east of Tasmania) alone during the time the sealers were there.[5]

Albatross hunters in the Auckland Islands. From a photo by Mr Dougal, 1888, Invercargill. Wood Engraving. State Library of Victoria.

One of the biggest impacts that humans have had on seabirds was on the penguins of Macquarie Island. Penguins form extremely large colonies on Macquarie Island and they were easy prey for sealers looking for further sources of oil. The Royal and King Penguins were the most commonly used as they formed the largest colonies. They were routinely rounded up, struck with clubs and thrown into the large purpose-made steam boilers (or digesters, as they are also known). As many as 2000 Royal Penguins were put through the boilers in a day, yielding approximately 40 gallons (~150 litres) of oil.[7] Many of the remains of these digesters can still be seen in the penguin colonies today, a lasting legacy of the slaughter that once occurred there.

King Penguins were nearly wiped out by this industry. By 1914 there was only one colony left on the island, representing a reduction of hundreds of thousands of birds.[7] By the early 1900s the culture of exploitation must have been changing, because concerns were starting to be voiced about the wholesale destruction of wildlife on Macquarie Island. The Tasmanian government was receiving letters requesting that the island be made a sanctuary from nearly every scientific society in the world and Sir Douglas Mawson was a vocal advocate for Macquarie Island to be turned into a sanctuary. As a result of these and other representations to the State government, in 1920 the commercial lease on Macquarie Island was not renewed. This marked the end of commercial exploitation of the wildlife on Macquarie Island.[7]

FERAL ANIMALS—THE SEALERS' LEGACY

WHEN SEALERS CAME TO Macquarie Island they brought many feral animals with them, including cats, rabbits, wekas (a flightless bird from New Zealand), sheep, goats, rats and mice. Whilst the sheep and the goats proved relatively easy to remove, the other feral animals have proved more persistent and some of them remain today. The cats in particular have exacted a heavy toll on the burrowing birds that are thought to have nested in large numbers all over the island.

A single cat can despatch hundreds of burrowing birds in a year, and when the cat population was at its peak, in the late 1980s or early 1990s, it is believed that there were several hundreds living on the island. Since the 1970s, vertebrate pest management has been carried out on Macquarie Island with successful eradication of the wekas in the late 1980s and the cats in 2000. The last cat was shot on Macquarie Island by a Tasmanian Parks and Wildlife Ranger, Mark Geyle, and between 2001 and 2004 there was a definite increase in the number of breeding burrowing petrels, particularly Grey Petrels.

In an attempt to control a rapidly expanding and destructive rabbit population, the *Myxoma* virus was introduced in the late 1970s. The introduction of this virus and its vector, a flea, resulted in a rapid drop in population numbers, and by 1985 effective control of rabbits was widespread throughout the island. *Myxoma* virus proved a successful control mechanism between 1985 and the mid-1990s, but since then a

Rabbits introduced by sealers in the early 1900s have wreaked destruction on the sensitive environment of Caroline Cove Amphitheatre on Macquarie Island.

The fertile green vegetation of Petrel Peak in 2001 (above) and the same slopes in 2004 (below), denuded following an increase in rabbit numbers.

disturbing increase in rabbit numbers has been documented. A series of warmer, drier winters and a reduction in the effectiveness of the *Myxoma* has resulted in numbers rising to the levels of the early 1980s. The reduction in rabbit numbers achieved over the first 20 years of using the *Myxoma* virus allowed much of this vegetation to return to abundance and coverage levels not seen for many years. Unfortunately, due to the increase in rabbit numbers and sustained grazing pressure over the last five years, many areas around Macquarie Island are now in a similar condition, or worse, to how they were prior to the release of *Myxoma* virus.

This is of particular concern around the coastal slopes and adjacent ridge-lines where the four species of albatross and several species of burrow-nesting petrel breed. Following the successful cat eradication, the numbers of burrowing petrels, particularly the Grey and Blue Petrel, increased significantly over just a few years. Unfortunately, due to the impact of heavy rabbit grazing in more recent times, many of these nesting sites are no longer suitable for these species and many will not return to breed there until the vegetation returns.

In light of the sustained increase in rabbit numbers, the deleterious effects on the vegetation and the subsequent effect on nesting seabirds, the Department of Primary Industries, Water and Environment, in conjunction with the Tasmanian Parks and Wildlife Service, is developing a plan to eradicate rabbits (and rodents) from Macquarie Island. Such a process is extremely expensive, given the isolated nature of the island and the widespread distribution of the rabbits, and at this stage there is no guarantee of funding. The impact of the current rabbit population on Macquarie Island is severe, and unless something is done soon, the implications for this once-pristine World Heritage Area are extremely serious.

PELAGIC TUNA FISHERIES

LONG-LINING FOR TUNA BEGAN in the late 1950s and early 1960s in the Atlantic, Pacific and Indian Oceans and continues to the present day. The Japanese distant-water fleet was one of the biggest, but fleets from Taiwan, Korea, Spain and Australasia have all fished in these waters for tuna. Throughout the 1960s the number of hooks set in these fisheries increased rapidly in all oceans, peaking at over 250 million hooks set by distant-water pelagic long-line fleets south of 30°S.[14]

It is difficult to be certain of exact hook numbers due to variability in how different fishing boats report their fishing effort; however, there is reasonably reliable information on the Japanese distant-water pelagic fleet fishing in waters south of 30°S.[14] Hook numbers of this fleet increased relatively quickly in the Pacific Ocean until they peaked in the 1970s. Over the next 15 years the number of hooks set each year varied between 20 and 50 million before starting a gradual decline in the 1990s, with fewer than 15 million hooks set in the late 1990s. Hook numbers peaked in the Indian Ocean in the mid-1980s at over 80 million hooks before declining through the mid to late 1980s and early 1990s. In the Atlantic Ocean the trend was slightly different: through the 1980s hook numbers increased rapidly to over 50 million set each year. Throughout the 1990s hook numbers varied, with peaks of over 50 million and troughs of under 20 million hooks set each year. These numbers illustrate the magnitude of long-line

Tuna is the target of many long-line fishing vessels.

Fisheries observers hold the limp body of an albatross recently drowned on one of the long-line fishing hooks. Photo: Graham Robertson.

fishing operations. Clearly, even if the catch rate of birds is low, when the magnitude and extent of these fisheries are considered (i.e. millions of hooks each year), the impact on albatrosses and other petrels can still be significant.

Many albatross populations around the world have been declining over the last three decades and most of these declines have been associated with long-line fisheries.[15] The number of hooks set, the spatial locations of the operations, the type of long-lining and the methods used to reduce seabird by-catch have varied a good deal over this period.[16,17] Albatross populations have fluctuated in response to these varying degrees of threat.[18,19]

The biological and ecological characteristics of albatrosses make them particularly vulnerable to any elevated death rates. For example, the onset of sexual maturity in albatrosses is late, clutches consist of a single egg and breeding success is low. Under natural conditions, however, adult albatrosses generally have a high survival rate and they live a long time. Artificially increased death rates, such as those wrought by long-line fishing, render populations particularly vulnerable. To ascertain trends in population numbers, population studies need to be long term.

All four species breeding on Macquarie Island are known to be caught and killed in southern hemisphere long-line fisheries for tuna, including those operating in Australian[16,20] and New Zealand waters.[21] The small size of the breeding populations of Wandering, Black-browed and Grey-headed Albatrosses on Macquarie Island make them particularly vulnerable. Shy Albatrosses have also been caught and killed on long-line fishing vessels,[20] and a satellite-tracking study conducted in the 1990s showed that their distribution overlapped significantly with that of Japanese long-liners fishing for tuna in Australian waters.[22]

In some areas of the world the threat to some albatross populations has been lessened by a reduction in the number of hooks set and by fishers adopting practices that minimise the likelihood of catching seabirds. Vessels can employ practices such as setting lines at night (albatrosses typically don't feed at night), reducing offal discharge, using thawed baits or weighted lines that sink before birds can pluck the bait from the surface, or streamer lines that scare birds away and thus reduce the risk of catching them. In addition, areas like the Australian Economic Exclusion Zone and waters under the jurisdiction of the Convention for the

These two Wandering Albatrosses have been hauled in, drowned on the end of long-line hooks, like so many are each year. Their mates will wait in vain for many days at the nest before abandoning it in search of a new partner. Photo: Rachael Alderman.

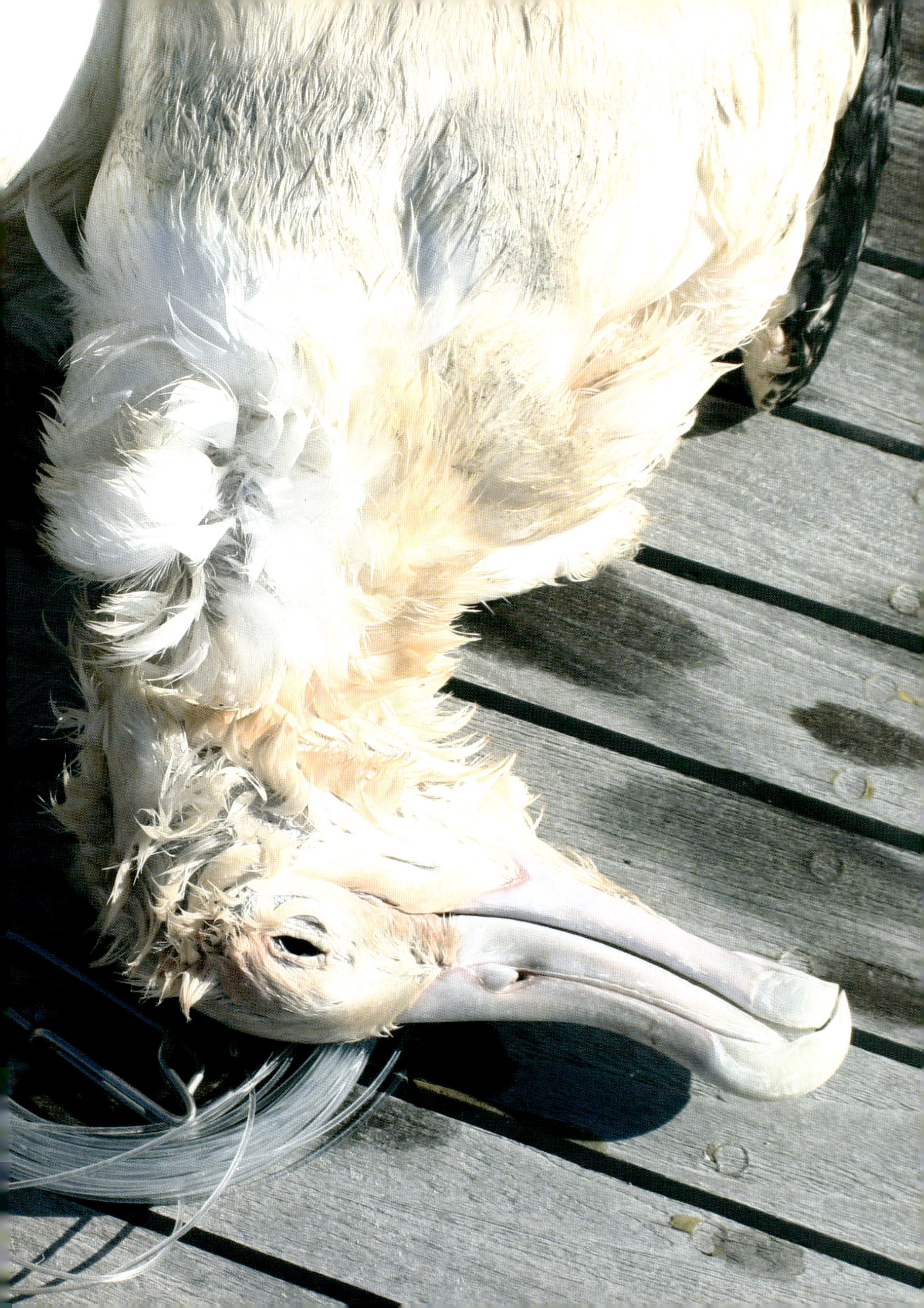

Conservation of Antarctic Marine Living Resources (CCAMLR) are strictly regulated: all vessels fishing in these waters must comply with a number of measures aimed at reducing seabird by-catch.

Despite these measures, many birds are still getting caught and killed. As the number of hooks set for tuna declines in some areas, those fishing for Patagonian Toothfish have increased in others, and there is a large, illegal fishery for this species operating in many of the more remote parts of the world,[14] particularly in the Southern Ocean where many of the albatross species breed and forage.

PATAGONIAN TOOTHFISH FISHERIES

THERE ARE SEVERAL OTHER long-line fisheries around the world that have impacts on seabirds, including those fishing for tuna-related species like Swordfish (*Xiphias gladius*), and other fish including Kinglip (*Hoplobrotula gnathopus*) and Hake (*Rexea solandri*). One of the largest and most rapidly expanding are the fisheries for Patagonian Toothfish. In contrast to long-lining for tuna, long-line vessels fishing for Patagonian Toothfish use demersal (bottom) long-line fishing techniques; however, the baited hooks still take time to sink once set, and present a similar threat to albatrosses.

Fisheries targeting Patagonian Toothfish began in the late 1980s and have increased dramatically over the last 20 years.[14] Originally developed by the Chileans, fisheries for this species now occur in many regions of the Southern Ocean, including the shelf and shelf-slope waters of the Antarctic continent, around many subantarctic islands and over seamounts. These areas correspond to those around which seabirds tend to feed, so there is often potential for by-catch.

Amongst the first observations of birds interacting with Patagonian Toothfish vessels were those undertaken by Greenpeace sailors on the *Gondwana* in the vicinity of South Georgia in March 1991.[23] They came across two Soviet vessels long-lining for Patagonian Toothfish and recorded six dead seabirds on three lines, including one Black-browed Albatross, one unidentified albatross and four White-chinned Petrels (*Procellaria aequinoctialis*). The number of hooks set by these fisheries in this area was estimated at around 5 million each year.[23] If a catch rate of this type was typical, it is possible that over 1000 albatrosses and 2000 smaller petrels were getting caught each year.[23]

The main albatross species that have been caught in Patagonian Toothfish fisheries include Black-browed and Grey-headed Albatrosses,[17] and to a lesser degree Wandering Albatrosses.[24] The most commonly caught seabird is the White-chinned Petrel; other seabirds caught include giant petrels and Grey Petrels.[17]

It can be hard to adopt practices that help reduce the by-catch of all seabirds as different species often behave quite differently.

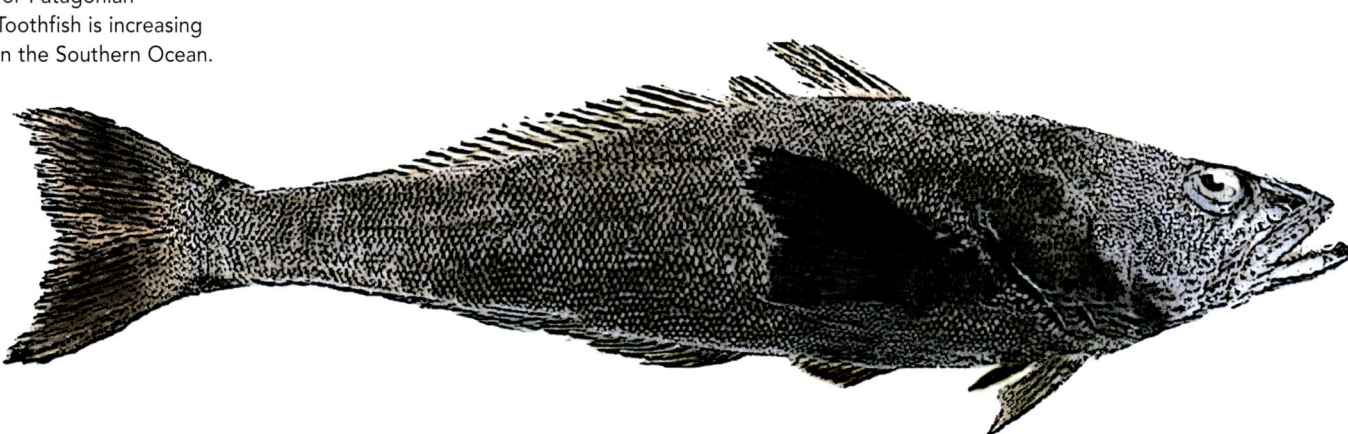

Long-line fishing for Patagonian Toothfish is increasing in the Southern Ocean.

For example, whilst setting lines at night may reduce the by-catch of albatrosses, it will not eliminate the by-catch of White-chinned Petrels as they typically forage in darkness.

There is also evidence of a skewed sex ratio in the bird by-catch of some Patagonian Toothfish fisheries, with males of several species being caught more frequently than females. This was found to be particularly true with the Grey-headed Albatross and White-chinned Petrels that have been caught around the Prince Edward Islands, a subantarctic group of islands under the jurisdiction of South Africa.[25] By-catch also varies seasonally, with higher catch rates occurring during the southern hemisphere (austral) summer around breeding islands. More birds are caught closer to their breeding sites than while at sea.[17] Obtaining this sort of information is extremely useful for both conservationists and fisheries managers, as seasonal or area closures can be introduced as a simple and effective means of reducing the by-catch of seabirds. However, while this has the potential to work well with licensed vessels in regulated waters, the proliferation of illegal and unregulated fishing activity means that by-catch of seabirds is still likely to be a significant issue. Some estimates of illegal fisheries operating around the Prince Edward Islands suggest that by-catch is as much as 20 times higher than in the regulated fishery.[26]

TRAWLING

LONG-LINE FISHING IS NOT the only threat to seabirds. Recent studies have revealed that trawling can also be responsible for the deaths of seabirds at sea. Seabirds interact with trawlers in a number of ways. Many seabirds have learnt to scavenge prey caught within the trawl nets during the net-hauling process. Some are injured or drown as they become entangled in the net or in lines that make up the trawl gear itself. Birds may also be injured or killed after colliding with trawl apparatus or by becoming stuck to lubricated cables and being dragged through trawl winches.[27]

There is now evidence from several trawl fisheries around the world indicating that trawl by-catch is a significant issue in subantarctic waters. By-catch levels appear to be related to offal discharges, with those fisheries that discharge more offal recording higher levels of mortality.[27] Given that offal discharge is prohibited in Australian subantarctic fisheries, it is likely that the level of by-catch in these fisheries is low. However, dirty water from fish processing can still lead to large numbers of birds around vessels in these fisheries. Reports of deaths in Australian fisheries have been low,[28] but further observations are required to clarify the interactions of seabirds and trawlers.

In a comprehensive review of threats facing seabirds in the Australasian region, it has been suggested that the only reliable way of ascertaining levels of seabird by-catch during trawl operations is to increase the number and coverage of fisheries observers on trawlers to focus on this issue.[27]

Long-line hooks loaded and ready for setting in the Patagonian Toothfish fishery. Each magazine of hooks is about 1.8 kilometres long and holds around 1300 hooks. Each vessel may set as many as 50 000 hooks. Photo: Graham Robertson

Churning up the water with their wildly flapping wings and squawking noisily, Wandering Albatrosses, Black-browed Albatrosses and giant petrels squabble over offal discarded from a long-line fishing vessel. Photo: Graham Robertson.

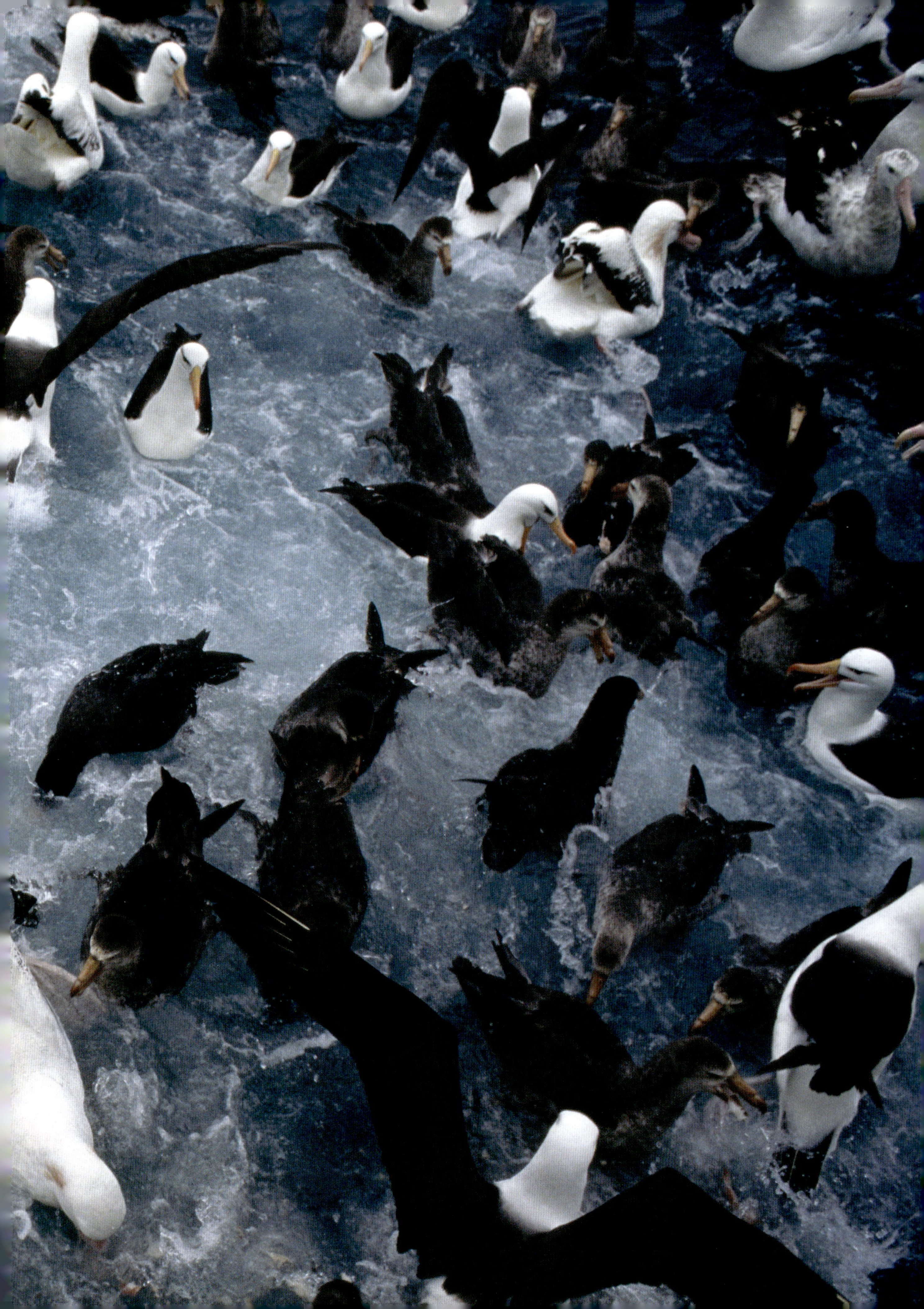

WANDERING ALBATROSS
Diomedea exulans

SHY ALBATROSS
Thalassarche cauta

BLACK-BROWED ALBATROSS
Thalassarche melanophrys

GREY-HEADED ALBATROSS
Thalassarche chrysostoma

LIGHT-MANTLED SOOTY ALBATROSS
Phoebetria palpebrata

THE ALBATROSSES

THIS CHAPTER DESCRIBES THE FIVE species of albatross that breed in Australian waters—four on Macquarie Island and one around Tasmania. Albatross species are difficult to distinguish in flight. Look for bill and feet colour and underwing edging.

WANDERING ALBATROSS
Diomedea exulans
LENGTH: 107–135 cm
WINGSPAN: 254–351 cm
WEIGHT: 6–10 kg
FEATURES: Huge, full-bodied albatross with extremely long wings and short, gently wedge-shaped tail. Iris brown, orbital ring blue or pink, bill white to pinkish. Legs and feet pinkish to bluish white with horn-coloured nails. Very similar in size and colouring to the Royal and Antipodean Albatrosses breeding in New Zealand.[1]

SHY ALBATROSS
Thalassarche cauta
LENGTH: 90–100 cm
WINGSPAN: 212–256 cm
WEIGHT: 3.4–4.4 kg
FEATURES: Medium to large albatross with a white crown forming a pronounced cap bordered by a narrow darker grey brow shading to lighter grey across the head. Grey on head is variable in deepness of colour and extent. Bill grey, grading to a more yellow tip. Iris dark brown. Legs and feet have bluish grey flesh. Very similar to the White-capped Albatross (*Thalassarche steadi*) breeding on New Zealand subantarctic islands.[1]

BLACK-BROWED ALBATROSS
Thalassarche melanophrys
LENGTH: 80–95 cm
WINGSPAN: 210–250 cm
WEIGHT: 3–5 kg
FEATURES: Similar in size and shape to the Grey-headed Albatross. Adult has distinct combination of white head with neat black brow. The bill is bright yellow–orange, with a reddish tip. There is a broad black leading edge on the underwing. Iris dark brown, legs and feet blue or light grey to white. Closely related and similar in appearance to the Campbell Albatross breeding on New Zealand subantarctic islands.[1]

GREY-HEADED ALBATROSS
Thalassarche chrysostoma
LENGTH: 71–85 cm
WINGSPAN: 180–205 cm
WEIGHT: 3.0–3.7 kg
FEATURES: Medium-sized albatross with distinctive combination of wholly grey head and black bill with narrow yellow stripes along the top of the bill becoming peach–red on front section of bill. Mostly white underwing with broad black leading edge. Iris is brown and legs and feet are white suffused with grey or pink. Superficially similar to the Buller's (*Thalassarche bulleri*) and the Yellow-nosed Albatross (*T. chlororynchus*).[1]

LIGHT-MANTLED SOOTY ALBATROSS
Phoebatria palpebrata
LENGTH: 80–90 cm
WINGSPAN: 180–220 cm
WEIGHT: 2.8–3.1 kg
FEATURES: Slender-bodied, dark-headed, grey and black albatross, with very long narrow wings. Characterised by long wedge-shaped tail and superb flying ability. Distinctive white crescent encircles eye, which has a brown iris. Dark slaty brown underwing with combination of blackish and pale white shafts in the tail feathers. Legs and feet mauve or greyish flesh. Similar to the closely related Sooty Albatross (*Phoebetria fusca*) that breeds on islands in the Indian Ocean.[1]

BILL SHAPE AT ACTUAL SIZE

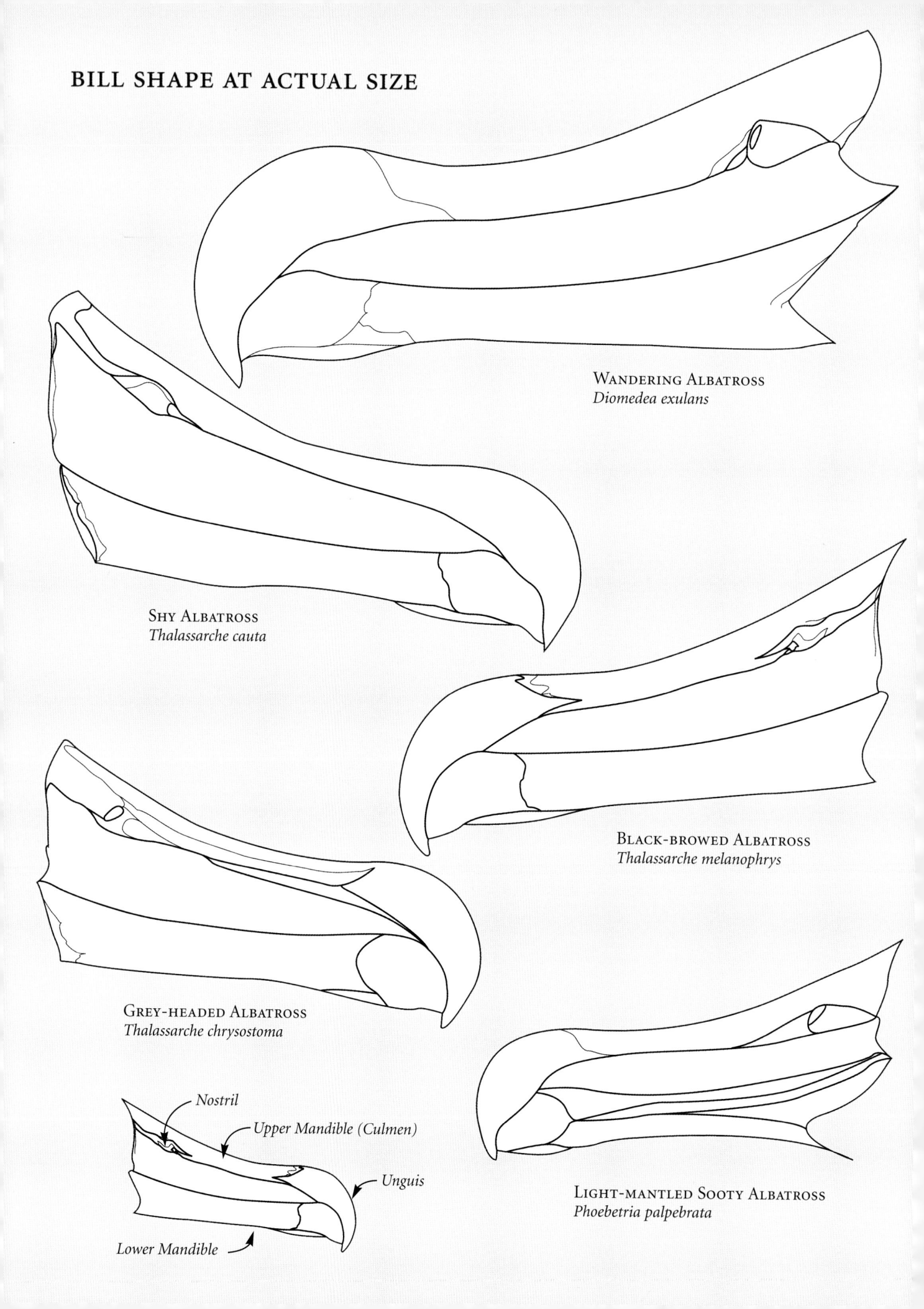

WANDERING ALBATROSS
Diomedea exulans

SHY ALBATROSS
Thalassarche cauta

BLACK-BROWED ALBATROSS
Thalassarche melanophrys

GREY-HEADED ALBATROSS
Thalassarche chrysostoma

Nostril

Upper Mandible (Culmen)

Unguis

Lower Mandible

LIGHT-MANTLED SOOTY ALBATROSS
Phoebetria palpebrata

FLIGHT PROFILES

Wandering Albatross shown at half size relative to other albatrosses.

Wandering Albatross
Diomedea exulans

Shy Albatross
Thalassarche cauta

Black-browed Albatross
Thalassarche melanophrys

Grey-headed Albatross
Thalassarche chrysostoma

Light-mantled Sooty Albatross
Phoebetria palpebrata

Wandering Albatrosses are constantly preening their feathers to maintain a sleek, groomed appearance and to keep themselves free of annoying parasites.

Wandering Albatross
Diomedea exulans

Breeding Sites
South Georgia, Iles Crozet, Iles Kerguelen, Marion Island, Prince Edward Islands, Macquarie Island

LIFE HISTORY

Wandering Albatrosses are the largest and probably the most visually striking of all albatrosses on Macquarie Island. Their breeding season begins in late November or early December, when the experienced males first return to land. It is not known if the pair communicates or interacts at sea, but not long after the male has made his first appearance the female arrives and they begin pair and nest-bonding. They will spend only a few days together at this first meeting, sporadically flying off short distances and returning to the island, and it is over these few days that the female's egg is fertilised. The female then departs the island and does not return for 10–14 days, while the male spends most of his time on the island continuing the nest preparations.

Even though pairs often return to the same nesting site, they usually build a new nest each year and it takes a lot of effort. Once the female returns, egg-laying is usually imminent and nest-building begins in earnest. One of the pair sits beside the nest pulling up clods of earth and grass and throwing these bits towards the nest, while the other picks them up and pats them into a large bowl-shaped nest that may end up being a metre across and 60 centimetres high. The sight of two Wandering Albatrosses working tirelessly over a two-day period to produce such an impressive structure with just their bills is surely one of the best examples of mates co-operating in the animal kingdom.

Once the nest is completed the female lays the egg and begins incubating. The egg is large: 13 centimetres long and 8 centimetres wide at its widest point and weighing up to 0.5 kilogram.[2] Once laid, the egg is manoeuvred up into the 'brood patch', which is a patch of downy skin on the underside of the bird that helps keep the egg

A young male Wandering Albatross makes his way up Petrel Peak, exhibiting the strange style of walking commonly referred to as 'vacuuming'. Banded as a juvenile on Macquarie Island, he is likely to be a visitor from another breeding location.

THE ALBATROSSES

at a warm and constant temperature. In some pairs this first shift may last up to a week, but usually the female lays and is relieved by her mate within a couple of days.

The female's first foraging trip once incubation begins is usually 7–21 days long.[3] During this time the male sits patiently on the nest, waiting for her to return. Once she returns, a changeover occurs and it is the female's turn to incubate while the male goes out foraging. This process continues for 79–80 days, which is how long it takes a Wandering Albatross egg to hatch. In the event of something happening to one of the pair on a foraging trip (for example, getting hooked and killed on a long-line), the mate will sit on the egg for up to 45 days before abandoning it.

Wandering Albatrosses have one of the strongest pair and nest bonds of all albatross species. A bird will sit and wait for a partner until almost all its body reserves are used up. Once an egg is abandoned it is quickly despatched by the watchful Subantarctic Skuas, predatory gull-like birds that live on Macquarie Island.

Not all eggs hatch. Between 1994 and 2001, Wandering Albatrosses on Macquarie Island had an average success rate of 65 per cent.[3] Most failures are due to infertile eggs, that is, for some reason the female's egg was not fertilised and so failed to hatch. This may be because one of the pair is infertile or because successful copulation has not occurred. One pair on Macquarie Island has laid and failed every year for 10 years in a row; in this case it is likely that one of the pair is infertile. This pair is exceptional as pairs usually split up after three successive failed breeding attempts.

If the egg does hatch successfully, the small chick emerges two to three days after first chipping a hole in the eggshell. These chicks weigh just a few hundred grams at birth and are carefully looked after by their parents. One or other of them remains constantly with the chick for the first 25–35 days of its life (known as the brood-guard stage). During this time the shifts on and off the nest shorten dramatically to just a few days.[3] This is because the chick requires more frequent feeding in its first few weeks of life. Albatrosses have a downy patch of skin on their underside known as a brood patch. Together with the surrounding feathers, the brood patch not only keeps the egg warm during incubation; it also protects the small chick from the elements. During the brood-guard period the chick spends much of its time asleep, waking

A young Wandering Albatross flies around the Caroline Cove Amphitheatre, Macquarie Island, having just returned to land after four years at sea.

LEFT:
Wandering Albatross (*Diomedea exulans* Linn.), 1884. Watercolour by H. C. Richter in Gould's *The Birds of Australia*. National Library of Australia.

BELOW:
Southern Ocean Wanderer, 2005. Oil on canvas. Peter Hall. Collection of H. & R. Whelan.

THE ALBATROSSES 81

Courtship rituals among albatrosses are elaborate and help seal the bond between lifetime partners. This pair of young Wandering Albatrosses is courting in the Caroline Cove Amphitheatre, Macquarie Island.

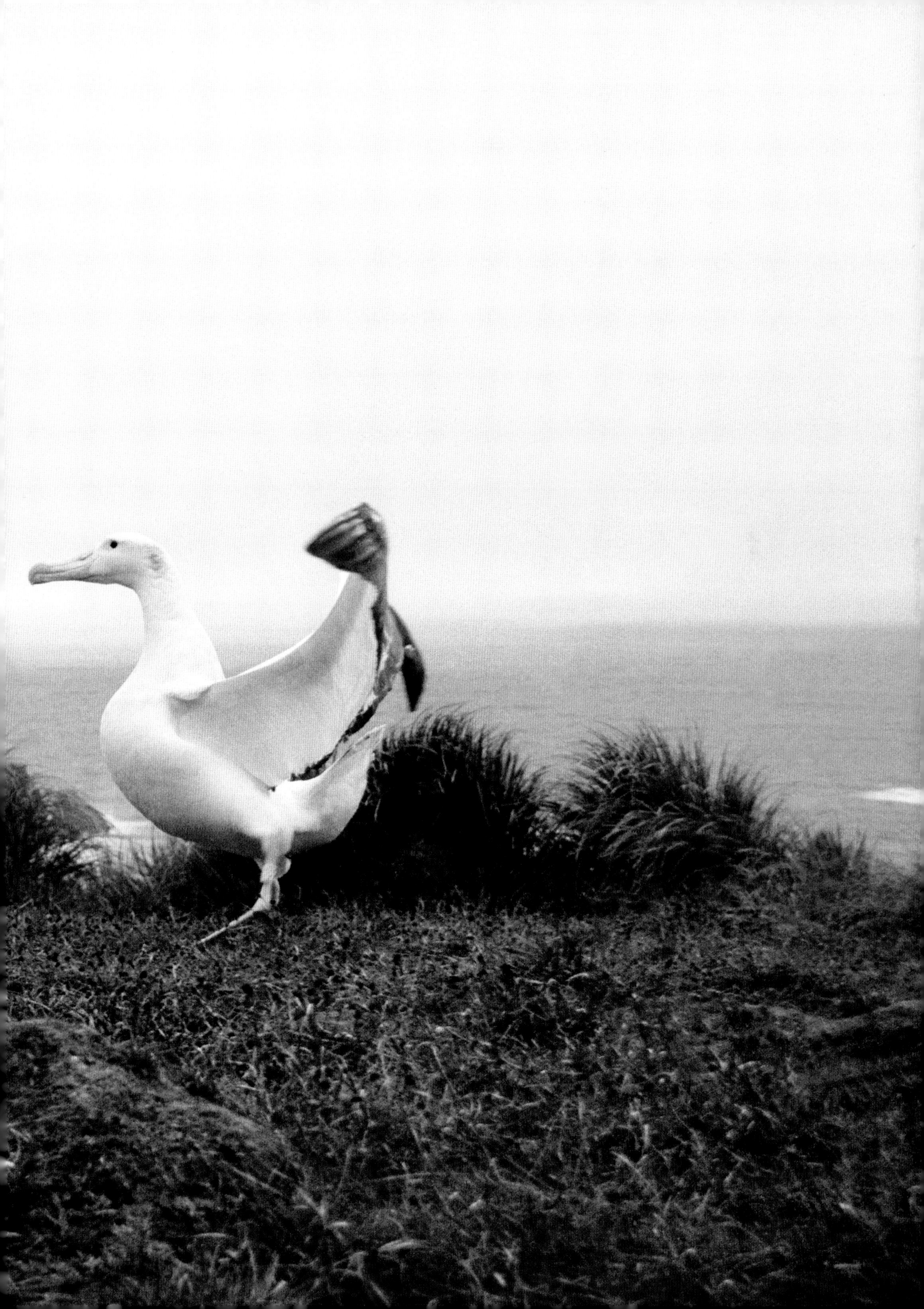

Wandering Albatrosses stretch their wings out to their full 3-metre extent during courtship. The stippled grey on the back of the closer bird identifies it as a young female.

only to eat or excrete. After 25–35 days the chick becomes too big for the brood patch and it is left alone for the first time. The sight of a young albatross chick at this point in its life is truly nature at its most vulnerable. The chick looks around at the world, exposed to the elements for the first time, its eyes bright with expectation, apprehension and anticipation. There is some fear in its eyes but also an aggressiveness that it has learnt from its parents in its first few weeks. Skuas and rabbits that come too close are snapped at and any kind of serious threat is treated to an immediate regurgitation of stomach oil, enough to dissuade most would-be predators from trying again.

However, there are very few natural predators on Macquarie Island, with skuas and giant petrels constituting the biggest threat. For this reason, and because of the strong pair and nest bonds of the parents, the success of chicks on Macquarie Island is very high, with success rates in a year usually above 90 per cent. Combined with the percentage of observed hatching success, this resulted in an average breeding success of 60 per cent throughout the 1990s.[3]

This is lower than is typically observed at other breeding locations, such as South Georgia and Iles Crozet. The reasons for this are not yet well understood. As there are so few pairs breeding on Macquarie Island, the failure of one or two pairs could be disproportionately lowering the breeding success rate.

Once left alone, the chick will sit on the nest for up to nine months, throughout the Macquarie Island winter, not moving off the nest through snow and squalls, freezing temperatures and winds often in excess of 40 knots. During this period the chicks are fed at regular intervals by the parents but, due to the long distances that they must travel, chicks may go without food for several weeks. Each parent usually brings in 500–1500 grams at each feeding event and this sustains the chick throughout the winter months.[4] During this time the chicks slowly lose their soft, fluffy down and gradually their fat turns into muscle. They reach their maximum weight a few weeks before leaving the island for the first time (fledging). As fledging approaches, they begin to exercise their developing wings.

Chicks usually fledge when their wings have

ABOVE:
A female Wandering Albatross incubating an egg. The dark pencilling in the plumage is characteristic of younger birds and it typically fades to white as they age.

RIGHT:
The black and white detail of Wandering Albatross feathers.

ABOVE:
Wandering Albatross on Display Hill, Macquarie Island, named for the tendency of young albatrosses to go there to 'display' themselves as part of courtship.

LEFT:
These young birds are in the very early stages of pair-bonding, having only recently returned to Macquarie Island for the first time after four years at sea.

THE ALBATROSSES 85

This parent patiently broods its two-week-old chick, but after 25–35 days, it will leave the chick alone for the first time. Wandering Albatross chicks are ready to fledge after nine months.

developed enough strength and it often happens by accident, as the chicks are exercising and a slightly bigger gust of wind lifts them up into the air. Once a chick has left the island it does not usually return to land for four to seven years, although one or two birds have been found on Macquarie Island at three years of age.

These first few years of life at sea must be difficult for the young albatrosses as they learn how to find food and survive in the oceanic domain. The survival of chicks at this time is relatively low, with an average of 45 per cent surviving to return to land after their first few years at sea.[5] This figure was calculated using over 40 years of breeding data and during the 1970s and 1980s, when long-lining was at its peak in the Indian Ocean, the survival of the young birds was probably much lower.[5]

Once the young Wandering Albatrosses return to the island they seek out mates and begin courting and trying to develop pair bonds. Young birds spend most of their time on land courting with other birds and, as they grow older, this courting and pair-bonding becomes more important as it is a critical step in successfully breeding and fledging a chick. Once a young pair has formed a bond they will return to the island just after the experienced breeders in December. The young pair will go through the motions of nest-building and pair-bonding but the female will not lay an egg and the couple depart the island sometime in January.

Young pairs will go through this process for two or three years before actually laying an egg. This role play ensures the strength of the bond between these long-lived birds and maximises the chances of them fledging a chick. No Wandering Albatrosses younger than eight years of age have been recorded laying eggs.

Occasionally, when an experienced breeding bird has lost its partner, new pair bonds may form more quickly with younger birds. For example, the mate of a 42-year-old male Wandering Albatross, who had had more than

The changes in the Wandering Albatross chick (opposite) at one month, three months, six months and nine months of age. The dark plumage will gradually lighten as the bird gets older.

THE ALBATROSSES 87

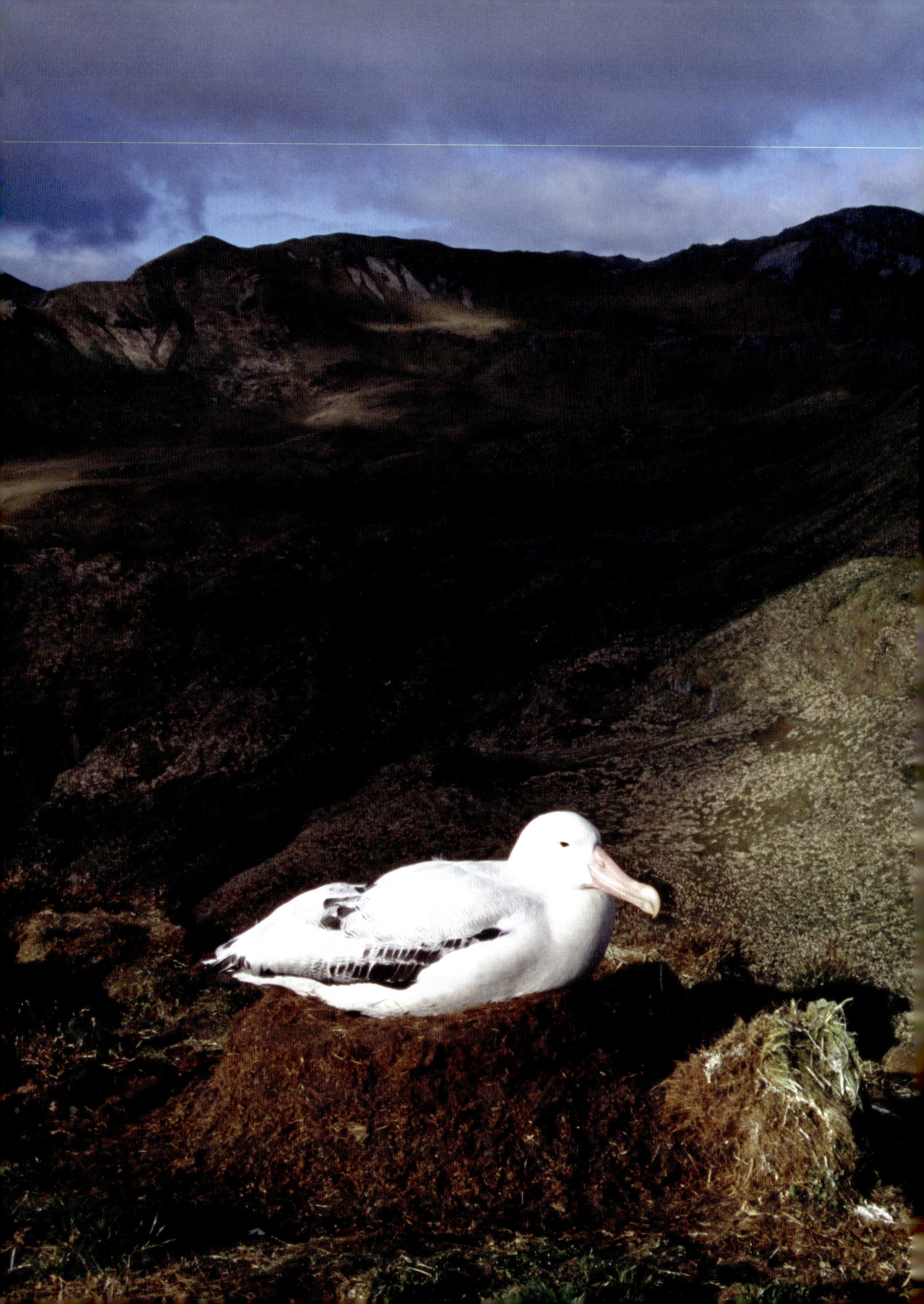

Illuminated by the sinking sun over Petrel Peak, this lone incubating Wandering Albatross sits it out. This nest was the largest of all the nests in its season and was built over three days by the pair working tirelessly together.

A young Wandering Albatross pair protects their newly established nesting site on Petrel Peak, with the imposing Mt Haswell in the background.

a dozen successful breeding attempts, did not return in 2002 to breed as expected. The male waited at the nest for most of the summer, but when it became clear that his partner was not going to return he began calling to and courting with unattached females. Having formed a pair bond with a younger female at the end of the summer, this new pair returned in December of 2003 and successfully laid and hatched a chick.

Wandering Albatrosses live for over 50 years, and several breeding birds on Macquarie Island are in their thirties or forties. Due to the time that it takes to fledge a chick, birds only breed every second year if successful and, once a chick has fledged successfully, the parents will not return to the island for 11–12 months. However, if a breeding attempt is unsuccessful, the pair may return the following year to try again. Annual survival of adult birds over the 50 years that records have been kept is relatively high with an average of 95 per cent.[5] This is relatively normal survival for adult albatrosses, but as with the younger birds, when long-line fisheries operations were at their peak during the 1960s and 1970s their death rate was probably much higher.

POPULATION TRENDS

The number of Wandering Albatrosses breeding on Macquarie Island has fluctuated considerably since the arrival of the sealers in the late 1800s. The discovery of over 100 Wandering Albatross skulls in Aurora Cave on the west coast of the island (remains left by shipwrecked sailors) suggests that numbers in the nineteenth century were significantly higher than the single breeding pair documented in 1912.[6] Following the departure of the sealers, the Wandering Albatross population slowly recovered until 1951 when there were at least 17 breeding pairs in total and, at its peak in the mid-1960s, the total breeding population was estimated at 44 pairs.[6] At this point the population extended right down the west coast of Macquarie Island with the highest densities in the north-west and south-west corners. The population then started to decline again in the late 1970s reaching another low point in the mid-1980s (less than five breeding pairs in total).[7] This population decline can be attributed to mortality due to long-line fishing operations.[5,7] Since then the population has gradually increased to approximately 19 breeding pairs in total in the late 1990s. It remained at this level between 1996 and 2003, with older breeders that disappear from the population being replaced by younger birds. The population is still precariously balanced and the deaths of only a few extra individuals a year would be enough to push this small and vulnerable population into decline and eventual extinction.

FEEDING HABITS

Of all the albatross species, Wandering Albatrosses are one of the widest ranging. Satellite-tracking studies from other subantarctic islands have shown that they can travel up to 13 000 kilometres in a single foraging trip,[8,9] and they travel right around the globe in the early years of their life.[10] Very little satellite-tracking on this species has been conducted on Macquarie Island birds but in recent years two individuals have been tracked each year.

The long trips that are often seen during incubation are not seen throughout the breeding cycle as parents during the brood-guard stage are likely to forage much closer to the island in order to satisfy the nutritional needs of their chick. Squid constitute most of the diet of this species, and while they can find squid close to Macquarie Island, albatrosses often travel vast distances in search of this prey, particularly during the incubation stage. Squid are highly mobile organisms that often congregate in large schools at 'fronts' within the Antarctic Circumpolar Current.[11] Squid only live one or two years and are thought to undergo massive die-offs after spawning.[12] Following the die-off, many species of squid float to the surface and, as Wandering Albatrosses usually feed on organisms close to the surface of the water, floating dead squid are a perfect source of food.

Females tend to forage in more northerly waters than males and as a result have historically been more at risk from long-line fishing operations. However, the relatively recent increase of long-liners fishing for Patagonian Toothfish in more southern waters means that the males from many populations are now also at risk.[13]

A Wandering Albatross egg can weigh up to half a kilogram and takes 79–80 days to incubate.

This Shy Albatross chick on the Mewstone has lost almost all of its down. Shed feathers and those from chicks that die don't go to waste—they are often woven into the next year's nest.

Shy Albatross
Thalassarche cauta

Breeding Sites
Albatross Island, Pedra Branca and the Mewstone (Tasmania)

LIFE HISTORY

THESE MEDIUM-SIZED ALBATROSSES are slightly bigger than Black-browed or Grey-headed Albatrosses but considerably smaller than Wandering Albatrosses. Shy Albatrosses were named by John Gould, who discovered the species in the 1800s at Recherche Bay in southern Tasmania. The difficulty that he had in catching this bird for his collection prompted him to call it the Shy Albatross.

Unlike other albatross species, they have very little time between the end of one breeding season (i.e. when the chick fledges) and the beginning of the next (preparing the nest). It is likely that, as with other species, the males arrive back first to secure the nesting site and prepare for the arrival of the female. Birds start arriving back in late June to early July to secure a nesting site, often in the exact location of the previous year's nest.

Shy Albatrosses nest in very dense colonies, often with less than a metre separating nests. In fact, the separation between nests can often be measured by how far a bird can reach while sitting on the nest. Due to the lack of soil in many of the breeding sites, nesting material is highly sought after and if nests were built too close together birds would spend all their time defending their nest from other birds that are trying to steal their nesting material. Even so, nests are regularly pilfered for materials by other birds and, if a nest is left for any length of time, its initial occupant may return to find its nest has been totally taken apart and built into its neighbours' nests.

Nesting material usually consists of what little soil there is. Due to the dryness of most breeding sites, much of the nest-building takes place after a rainy period. On nests that are built close to the edges of the colony, grasses are also sometimes used in the construction of the nest.

Shy Albatrosses do not appear to be too picky about the nesting materials they use, and in many cases feathers from chicks that have died

An adult Shy Albatross on Albatross Island guards the material of an empty nest to prevent pilfering by other birds before egg-laying. Around 5000 pairs of Shy Albatrosses breed on this island each year.

the previous year are built into the new nest the following year.

Nest-building begins in earnest once both birds have returned. Nests vary considerably in shape and size, depending on the available materials. In some areas, particularly on the Mewstone where there is very little available nesting material, eggs are laid on bare rock or in a shallow, rocky depression. Due to the proximity of other birds, and the often aggressive interactions that ensue, many of these eggs fail. At the other end of the spectrum, where there is available nesting material and a nest has been used for several consecutive years, the cumulative nest-building effort results in a nest bowl of up to 40 centimetres in height, providing a much more secure environment in which to incubate the egg and eventually raise the chick.

Eggs are generally laid throughout August and most have been laid by early September. The male and female alternate relatively short shifts of a day or two, but occasionally they spend several days at sea in one trip before relieving their partner.[14,15] Common causes of egg failure probably include desertion of the nest by an incubating bird that is getting too hungry, eggs falling out of poorly made or small nests and eggs being cracked accidentally by one of the pair during a changeover. Occasionally eggs don't hatch, and in these cases it is likely that the female was not fertilised or one of the pair was infertile.

Those eggs that do survive have generally hatched by mid-December. At this stage shifts become slightly shorter and changeovers occur more frequently. In a comprehensive study on the feeding of Shy Albatross chicks at Albatross Island, April Hedd and co-workers discovered some very interesting facts.[16] These researchers used automatic weighing nests to remotely study chick feeding over three successive seasons. Automatic weighing nests are made up of a fibreglass nest bowl on top of a delicate weighing platform. These are put under the chick in place of the original nest and the weight is recorded automatically every 10 minutes. They found that chicks received relatively small meals that weighed on average 372 grams. Chicks were generally left alone for the first time at around two to three weeks of age, and their meal size increased as they got older. On average, chicks were fed once a day and most feeding took place at the beginning and end of each day. During their four months or so on the nest, chicks were fed around 40 kilograms of food for an overall weight gain of about 4–5 kilograms.[16]

Survival of Shy Albatross chicks on the nest is extremely variable, both from year to year and between colonies. Chicks die for a variety of reasons, and in many cases it is impossible to identify a definite cause. In some cases the breeding environment gets very hot and on warm days the chicks pant rapidly in an attempt to lose excess heat. The high temperatures may be too much for some chicks, particularly if they

Two pairs of Shy Albatrosses on Albatross Island sit together before breeding. Spending time together, often preening each other, reinforces the pair bond that is vital for successful breeding.

ABOVE:
Shy Albatross (*Diomedea cauta*), 1884. Watercolour by H. C. Richter in Gould's *The Birds of Australia*. National Library of Australia.

RIGHT:
This Shy Albatross has spawned an albino chick. Albino chicks are rare—only one or two are seen each year—and they can be identified by the pure white plumage and pink feet and bill. Interestingly, researchers have not yet spotted an adult albino.

THE ALBATROSSES 95

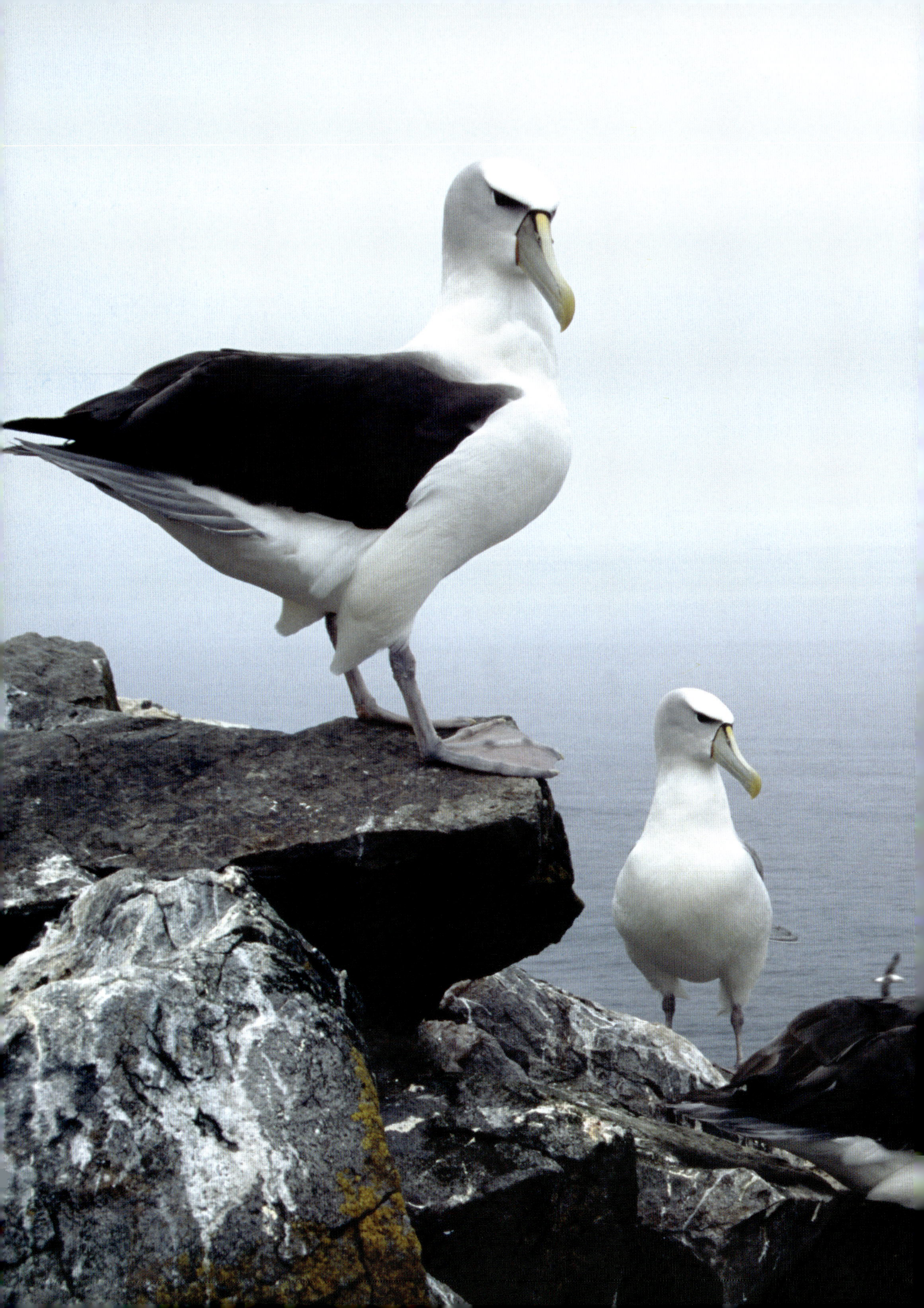

Congregated on top of the Mewstone, these Shy Albatrosses are resting from a recent foraging trip. Gatherings of this nature allow potential mates to be found and pair bonds to be formed.

ABOVE:
A Shy Albatross prepares for take off from the Mewstone to feed in the shallow waters close to the continental shelf.

RIGHT:
The slightly ruffled plumage of this Shy Albatross indicates that it is just weeks away from fledging.

98 ALBATROSS

are underfed or malnourished. There is also a disease that appears to strike some chicks on Albatross Island. In a bad year this can wipe out over 90 per cent of the chicks hatched in some colonies. The death of a parent early in the feeding regime will also usually result in the death of a chick as it is rare that a single parent can provide sufficient food for a chick throughout the time it is dependent on its parents.

Those Shy Albatross chicks that do survive normally reach a peak weight of about 5 kilograms and fledge around 110–130 days after hatching, usually in late April and May.[16] Once they leave the island chicks are thought to travel widely, with some earlier studies showing that they may reach as far as southern Africa.[17] The fate of the Shy Albatross chick after fledging was in the spotlight in 2004 with the advent of the Big Bird Race, where people were able to bet on the travelling speed and direction of some 18 Shy Albatross fledglings.

Shy Albatross chicks return to land for the first time at around two to six years of age. Generally they return to the island from which they fledged. The young birds spend most of their time on the periphery of the colonies, learning how to interact with other birds on land. These interactions involve courting and building 'play nests', which are good practice for forming pair bonds and, eventually, successful breeding.

POPULATION TRENDS

LITTLE INFORMATION HAS BEEN published on the trends in Shy Albatross population numbers. We know that the population on Albatross Island was nearly wiped out by feather hunters. Since the cessation of this practice, the population on the island has slowly recovered and it is currently estimated to be at around 5000 breeding pairs.[18] The total population on the three Tasmanian breeding islands is estimated at around 12 000–13 000 breeding pairs.[18] The way in which the Tasmanian populations are monitored has been reviewed. Improved technology now provides higher resolution in aerial photographs which in turn provides more accurate estimates of population numbers.

During the 1980s and 1990s Shy Albatrosses were caught and killed by long-line fishing vessels in Tasmanian and Australian waters,[19] and inevitably this extra mortality would have impacted on breeding numbers. Over the last 25 years data on population numbers of Shy Albatrosses has been collected and analysis is currently underway to clarify the population trends over this period.

A Shy Albatross collects mud and vegetation for nest-building in the Main Colony, Albatross Island. Young birds such as this are forced to build their new nests on the edge of the colony, away from more established pairs.

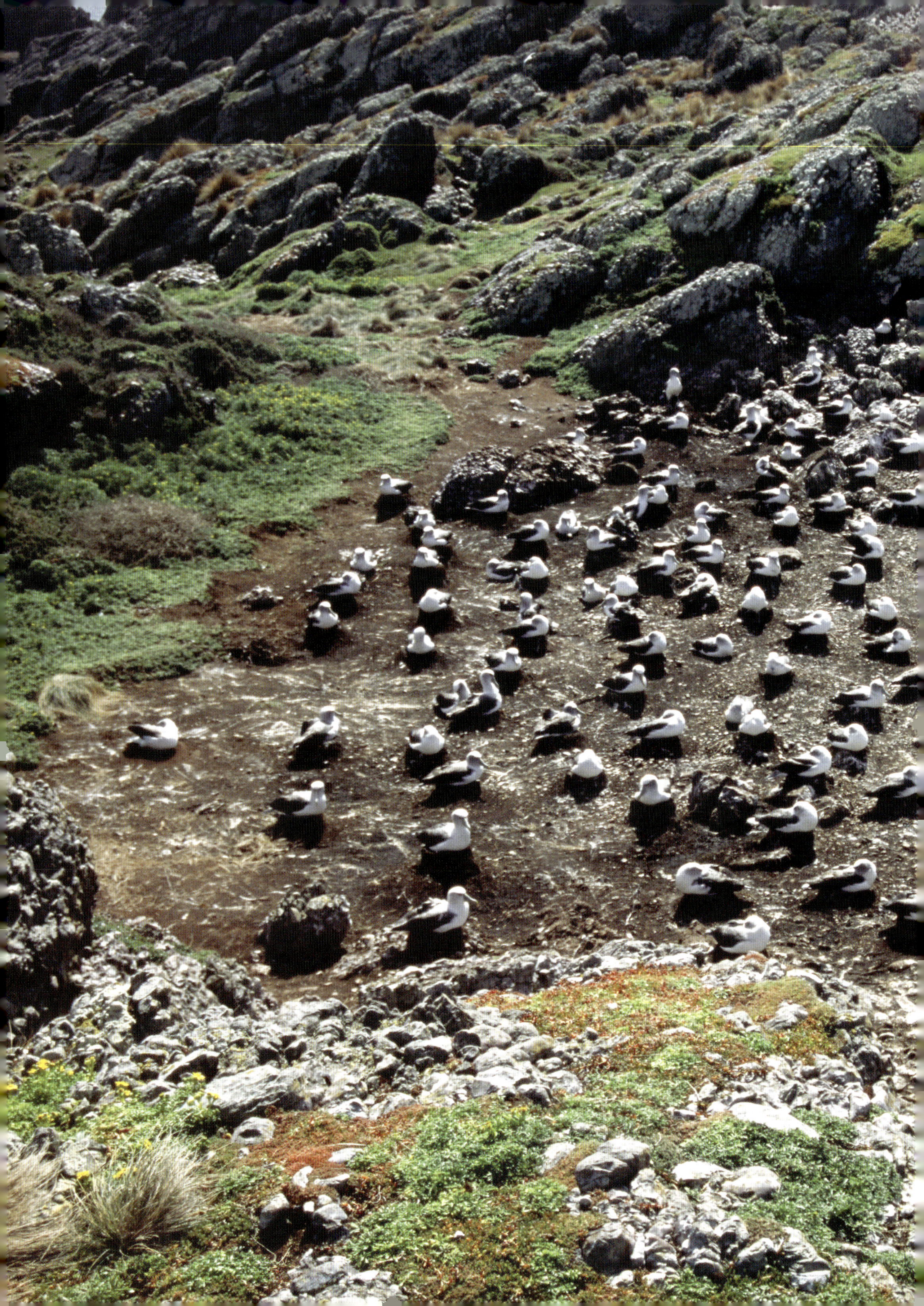

At South Colony on Albatross Island the breeding season has begun. Suitable nesting sites are limited but birds space themselves out far enough from their neighbours to prevent them reaching out to pilfer valuable nesting material.

An adult Shy Albatross on the edge of the crowded Main Colony unfolds its 2.5-metre wings to stretch.

FEEDING HABITS

Shy Albatrosses typically feed in relatively shallow waters, generally close to or on the margins of the continental shelf. Reflecting these feeding habits are the Shy Albatrosses' choice of breeding islands. Around Tasmania every one of these lies within 100 kilometres of the continental shelf.

During the 1990s breeding birds were satellite-tracked from the Albatross Island and Pedra Branca colonies during incubation and early chick-rearing.[14,15] These birds fed in relatively local waters, either over the continental shelf or along the shelf break but they never crossed into oceanic waters beyond.

The highest density of foraging birds from Albatross Island was found over a large area about 70 kilometres to the west of the island, near the shelf area. This foraging area shrank during early chick-rearing, with the highest density of birds shifting to just 9 kilometres to the west of the island.[14] Although during chick-rearing the foraging area diminished, there was still considerable overlap in foraging areas during incubation and early chick-rearing. Breeding birds from Pedra Branca ranged over a comparatively small area, foraging to the east or south-east of Pedra Branca towards the continental shelf.[14,15] As the breeding season progressed, foraging trips from all breeding islands lengthened in a similar way to those of other albatross species. The trips were longest during incubation and shortened as hatching approached, reducing still further through the early chick-rearing period.[14,15]

Shy albatrosses largely feed on squid and small fish.[20] They are thought to be more predators than scavengers and much of their prey is captured alive, at the surface, during the day.[20] Although primarily feeding on prey at the surface, this species has also been known to dive for its food, and dives to depths of over 7 metres have been recorded.[21]

Satellite tracking studies continue on the Shy Albatrosses to examine in more detail their distribution during their first year at sea, as well as during the non-breeding season.

ABOVE: A Shy Albatross lifts its wings in preparation for takeoff from the Mewstone with the hazy outline of Maatsuyker Island in the background.

LEFT: Bill-tapping is common in the process of pair-bonding and courting among Shy Albatrosses. This pair courts in the late afternoon light on Albatross Island.

THE ALBATROSSES 103

The prominent black brows from which the species gets its name can be clearly seen as these Black-browed Abatrosses perform bill-tapping as part of pair-bonding in the Main Colony, Macquarie Island.

Black-browed Albatross

Thalassarche melanophrys

Breeding Sites
Falkland Islands, South Georgia, Iles Crozet, Iles Kerguelen, Marion Island, Heard Island, Macquarie Island, Campbell Island, Antipodes Islands, the Snares, Chilean offshore islands

LIFE HISTORY

Black-browed Albatrosses are relatively small albatrosses that breed in low numbers on Macquarie Island. They generally breed every year regardless of prior breeding success. They are quite vocal birds and they tend to be more antagonistic towards each other than the other species breeding on Macquarie Island. Perhaps they show a different temperament because they nest in larger groups or colonies than other species breeding on Macquarie Island. Nevertheless, as with other species, it is impossible to generalise because, even in the midst of 15 or more breeding birds on nests less than a metre apart from their neighbours, quiet birds with extremely mellow temperaments can still be found.

There are four small, distinct colonies on Macquarie Island, the largest numbering 15–20 nests each year and the smallest two to five nests each year. The remaining breeding pairs nest singly, often interspersed among the Grey-headed Albatrosses on the steep tussocked slopes of Petrel Peak, the hub of albatross breeding on Macquarie Island. There are two small islets approximately 37 kilometres to the south of Macquarie Island (Bishop and Clerk Islets) where there are thought to be over 100 breeding pairs.[22] Black-browed Albatrosses used to breed on North Head in the north of Macquarie Island, but that colony died out through the 1960s and 1970s.

Experienced breeders first return to Macquarie Island in late August. As with other species of albatross, it is not known if the pairs communicate at sea but the males tend to arrive first, stake out their nest site and wait for the females. Males and females can often be distinguished by slight differences in their bill and head size, but these differences are often not

A Black-browed Albatross broods its small 10-day-old chick. Approximately 40–45 pairs of Black-browed Albatrosses breed each year on Macquarie Island. They usually nest in small groups but this nest is on the slopes of Petrel Peak, away from most others.

Black-browed Albatross (*Diomedea melanophrys* Temm.), 1884. Watercolour by H. C. Richter in Gould's *The Birds of Australia*. National Library of Australia.

pronounced and a practised eye is usually required to tell them apart. Following the first meeting the pair spend one to six days pair- and nest-bonding and copulating before the female leaves the nest to stock up on energy reserves for the egg-laying event.

The male spends most of this time building the nest and guarding it from other pairs looking for a good spot. Due to a scarcity of nesting materials on the rocky ground that the Black-browed Albatrosses often nest on, nests are often pulled apart by rival pairs in an attempt to get enough materials to build their own nest. Occasionally fights between individual birds break out; sometimes opponents locked together will tumble through the breeding colony disrupting sitting birds and creating a flurry of excitement. At these sites, nests are composed mainly of mud and guano, and painstakingly built up over a few days by the pair working together, one throwing the mud towards the nest and the other patting it into shape. The nests of Black-browed Albatrosses are smaller than those of Wandering Albatrosses, measuring about 30–40 centimetres across and 20–50 centimetres high. Nevertheless, they are still impressive structures, given the size of the bird. The walls are high and incredibly solid and the whole structure is very stable and may last for several years. Little wonder then that Black-browed Albatrosses often return to the same nesting site each year.

The egg is laid by the female between late September and early November about 7–14 days after copulation. The female usually does a short first shift and is relieved by the male after a few days. The incubation period lasts 66–72 days and the average shift lengths range from three days in early incubation to four days in late incubation.[3]

The hatching success of Black-browed Albatrosses is generally the lowest of all species on Macquarie Island. Through the mid to late 1990s it averaged just 54 per cent.[3] This comparatively low success rate may reflect a weaker pair bond, as birds tend to form pairs relatively quickly and at a young age. The annual breeding nature of this species (regardless of prior breeding success) and the short foraging trips also means that there is little time for consolidating and renewing that pair and nest bond at the beginning and throughout the season. Consequently, partners that are incubating on the nest sometimes show a greater tendency to abandon the egg if their partner undertakes a long or longer-than-usual foraging trip. These longer foraging trips may be the result of fluctuations in the prey items (such as squid, krill and small fish) in their normal foraging grounds.

Predators on the island may have some influence on the survival rates of young chicks too. Once hatched, the chick needs regular feeding and the foraging trips become even shorter, allowing the chick to be fed every one or two days on average. Black-browed Albatross chicks are occasionally left alone for the first time at the tender age of just 12 days and at this time they are extremely vulnerable. However, with tenacity belying their small size they are generally able to fend of the skuas, which are their only real threat. Before cats were eradicated from Macquarie Island it is possible that they were responsible for the deaths of some small albatross chicks, but there was little evidence of this and it seems that the defensive regurgitant response was usually enough to deter most would-be feline predators.

After it has been left alone for the first time, the chick remains on the nest for another three to four months before fledging in late April or May. As with the other species, chick success is typically higher than egg-hatching success for Black-browed Albatrosses on Macquarie Island, with 80–90 per cent surviving on average. Overall, the average breeding success through the mid to late 1990s was just under 50 per cent.[3]

The parents continue to feed the chick regularly over the three- or four-month period between being left alone and fledging, and the chick rarely goes for more than a week without food. As the chick gets older the frequency of the feeds decreases and, after reaching its peak weight a few weeks before fledging, its weight declines to levels similar to that of adult birds. Black-browed Albatross chicks become quite mobile in the weeks prior to fledging, and will walk around the colony, sitting on empty nests,

It's rare to see Black-browed and Grey-headed Albatrosses nesting in close proximity to each other, as they are here on West Rock, Macquarie Island.

This Black-browed Albatross keeps her small chick snug deep within the bowl of a large nest. These southern slopes of Petrel Peak are constantly blasted by icy ocean winds and only tufted tussocks of grasses can withstand the harsh climate and rugged terrain.

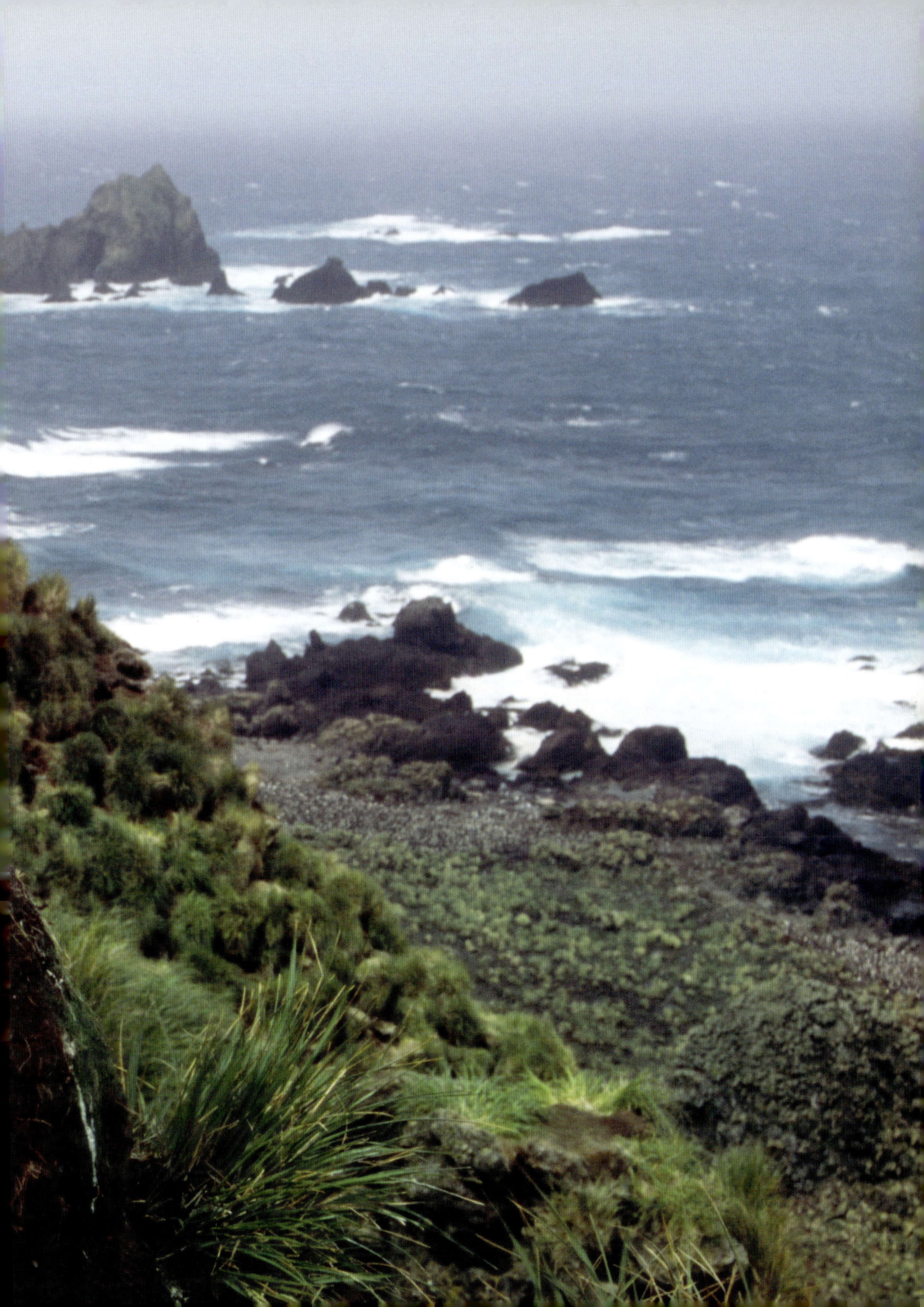

ABOVE:
The turbulent seas around Macquarie Island are fertile foraging grounds for squid, krill and small fish.

RIGHT:
A Black-browed Albatross prepares to take flight from its recently built nest. At the bottom of the slope hundreds of penguins are nesting.

and even try to elicit food from other parents.

Once the chick has fledged it does not return to Macquarie Island for several years. The youngest bird resighted to date was four years of age. There is little information on where these young birds from Macquarie Island go in these first few years of life but it is likely that they spend all their time at sea. As all chicks are banded with metal and colour bands, birds can occasionally be identified at sea and some juvenile Black-browed Albatrosses have been resighted off the south-east coast of Australia. The survival rate of young Black-browed Albatrosses is comparatively high, estimated at close to 60 per cent during the first four to six years of life that they spend at sea.[23]

When they return to Macquarie Island, the young birds spend most of their time interacting with other juvenile birds and, like all albatrosses, they attempt to form pair bonds. Unlike most Wandering Albatrosses, Black-browed Albatrosses form pair bonds relatively quickly and have been known to form a pair bond and breed in the same year. However, this is somewhat unusual and pair bonds are generally formed in one season with breeding occurring in the next. Black-browed Albatrosses have been recorded breeding as young as six years of age on Macquarie Island, but usually breeding does not begin until birds are seven years old.

Most Black-browed Albatrosses breed every year regardless of prior breeding success, with only a relatively small percentage (around 15 per cent) deferring breeding for one or two years.[24] This deferral may be in response to the birds not obtaining enough food over the winter months and consequently being in inadequate condition to see through successful breeding. The small proportion of birds that defer breeding also suggests that other breeding birds may be breeding in inadequate body condition, and this may also contribute to the comparatively low breeding success observed in this species.

Black-browed Albatrosses are one of the shortest lived species on Macquarie Island. This is probably in part attributable to the frequency of their breeding attempts. Breeding is an energetically expensive exercise and few Black-browed Albatrosses have been recorded as living over 35 years of age on Macquarie Island. The average annual survival rate of adult Black-browed Albatrosses from Macquarie Island over the last three decades is estimated at 90–92 per cent, while survival estimates for juvenile birds during their first five to seven years at sea are considerably lower, at around 60 per cent.[23]

POPULATION TRENDS
PRIOR TO THE BEGINNING of the long-term albatross study initiated by the Nature Conservation Branch in 1994, there was little information on breeding numbers of this species on Macquarie Island. Geof Copson, working for the Tasmanian Parks and Wildlife Service in the 1970s and 1980s, published his counts of breeding birds between 1977 and 1983.[25]

These numbers were somewhat lower than those of the late 1990s but the number of chicks that fledged in the two time periods was similar. Although chick counts are not always a good indication of population numbers, it is likely that the Black-browed Albatross population has remained steady since the 1970s.[25] The differences in the results of the studies are probably accounted for by the differences in methods used and the timing of the counts.

Old biological logbooks on Macquarie Island

On the lower slopes of Petrel Peak this three-week-old Black-browed Albatross sits up and takes notice of the world. In the background hundreds of penguins are gathered, as is their habit along this south coast of Macquarie Island during the summer months.

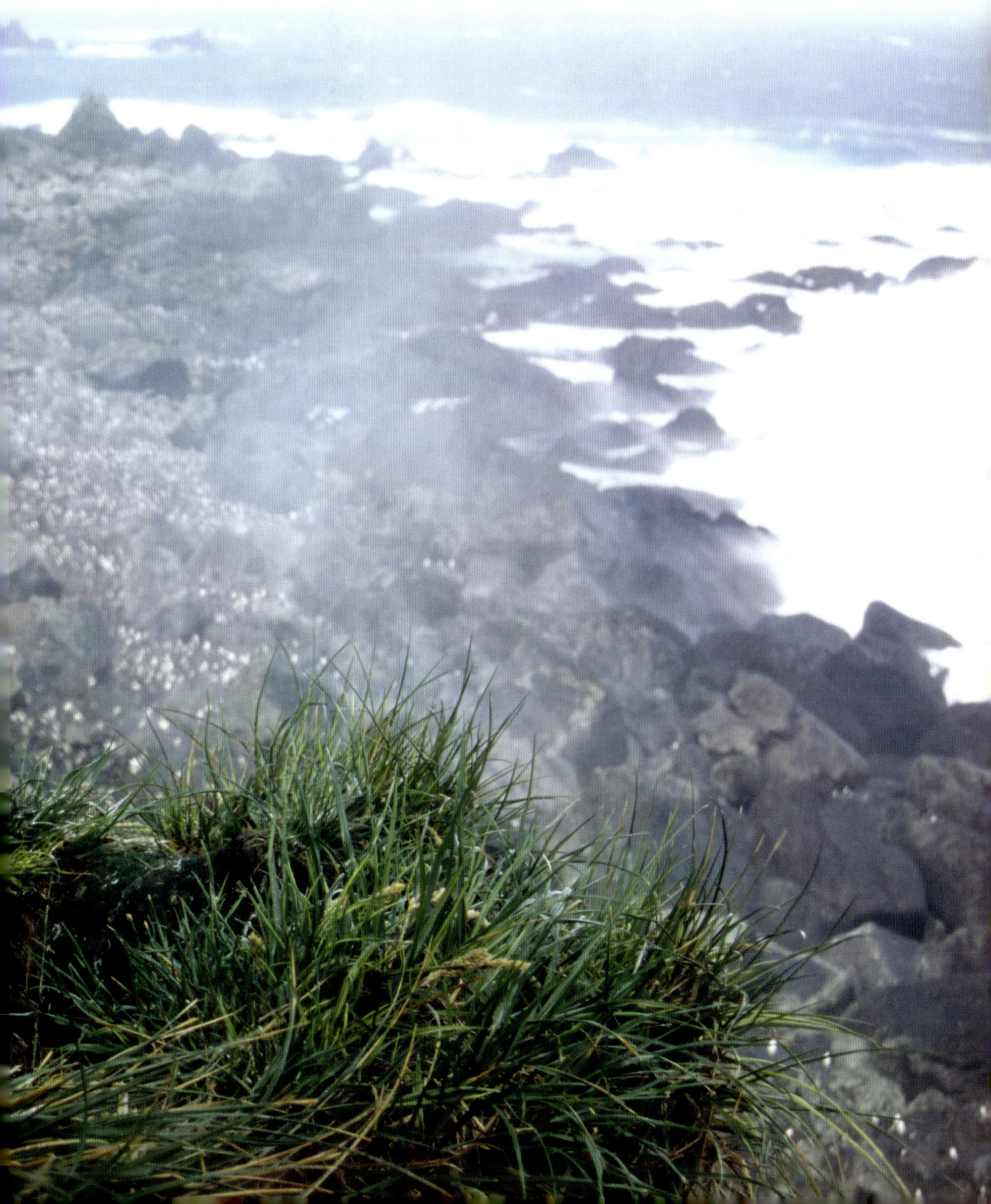

RIGHT:
Black-browed Albatrosses build solid, high-walled nests that may last for several years, and pairs will often return to the same nest each year.

BELOW:
A Black-browed Albatross glides in to land on Macquarie Island above raucous colonies of Royal and Rockhopper Penguins nesting on the rocky beach below.

A Black-browed Albatross makes a noisy return from foraging as it lands on Petral Peak.

provided some further information on the historical breeding numbers of the Black-browed Albatrosses there. There was a small breeding population of Black-browed Albatrosses on North Head that declined from 31 breeding pairs in the 1950s to fewer than three breeding pairs in the late 1970s.[25] Most of this decline occurred in the 1960s and 1970s and, in the absence of detailed population data, it is difficult to ascertain the causes of such a decline. It is possible that the decline was a result of human disturbance as the colony is very close to the main ANARE station, but given the lack of data it is impossible to be sure. In contrast, there is some evidence that the colony on Bishop and Clerk Islets has increased since the 1960s. A 1968 report estimated 25+ pairs nesting there[26] and a study in the 1970s estimated that there were at least 44 pairs nesting on these islets.[27] These estimates were obtained from ship-based observations and their accuracy is questionable. An estimate of 140 pairs[22] was made during a ground-based count in March 1992.

FEEDING HABITS

THE UNDERLYING NATURE OF THE ocean floor (bathymetry) appears to be one of the most important factors influencing where Black-browed Albatrosses from Macquarie Island feed. Most birds concentrate their efforts around the ridge complex to the north and south of Macquarie Island (see map page 116). Even though there is no shelf edge as such around Macquarie Island, the water movement around the ridge complex to the north and south of the island is likely to bring a lot of suitable prey up to the surface. Antarctic Fur Seals and Subantarctic Fur Seals breeding on Macquarie Island also spend most of their foraging time around the ridge to the north of the island.[28] In fact the main foraging areas of the Black-browed Albatrosses and these fur seal species are almost identical, suggesting that the waters in this area are highly productive.

Although most of the foraging effort is concentrated close to the island itself, several longer trips are made, most notably one into Antarctic waters and another that reached the south-eastern coast of Australia (see map below).[29] These two trips, each covering thousands of kilometres, clearly show that Black-browed Albatrosses from Macquarie Island do not restrict themselves to waters close to their breeding site. Presumably they undertake these long trips in order to find food they can't find close to Macquarie Island.

The Antarctic shelf waters and those off the Australian continental shelf are rich in chlorophyll A. Chlorophyll A can be measured by satellite imagery and represents the amount of phytoplankton in the water. Phytoplankton, and the small animals that feed on it (known as zooplankton), are the basic building blocks of all oceanic food chains.

High levels of chlorophyll A are often found at continental shelves where the movement of water into shallower regions creates upwellings. Other oceanic foraging areas of this species appear to be related to the Polar Front where warmer salty water meets cold water from around Antarctica. These areas are also likely to concentrate the food of this species.

Foraging areas of Black-browed Albatrosses from Macquarie Island during the breeding season.[29,34]

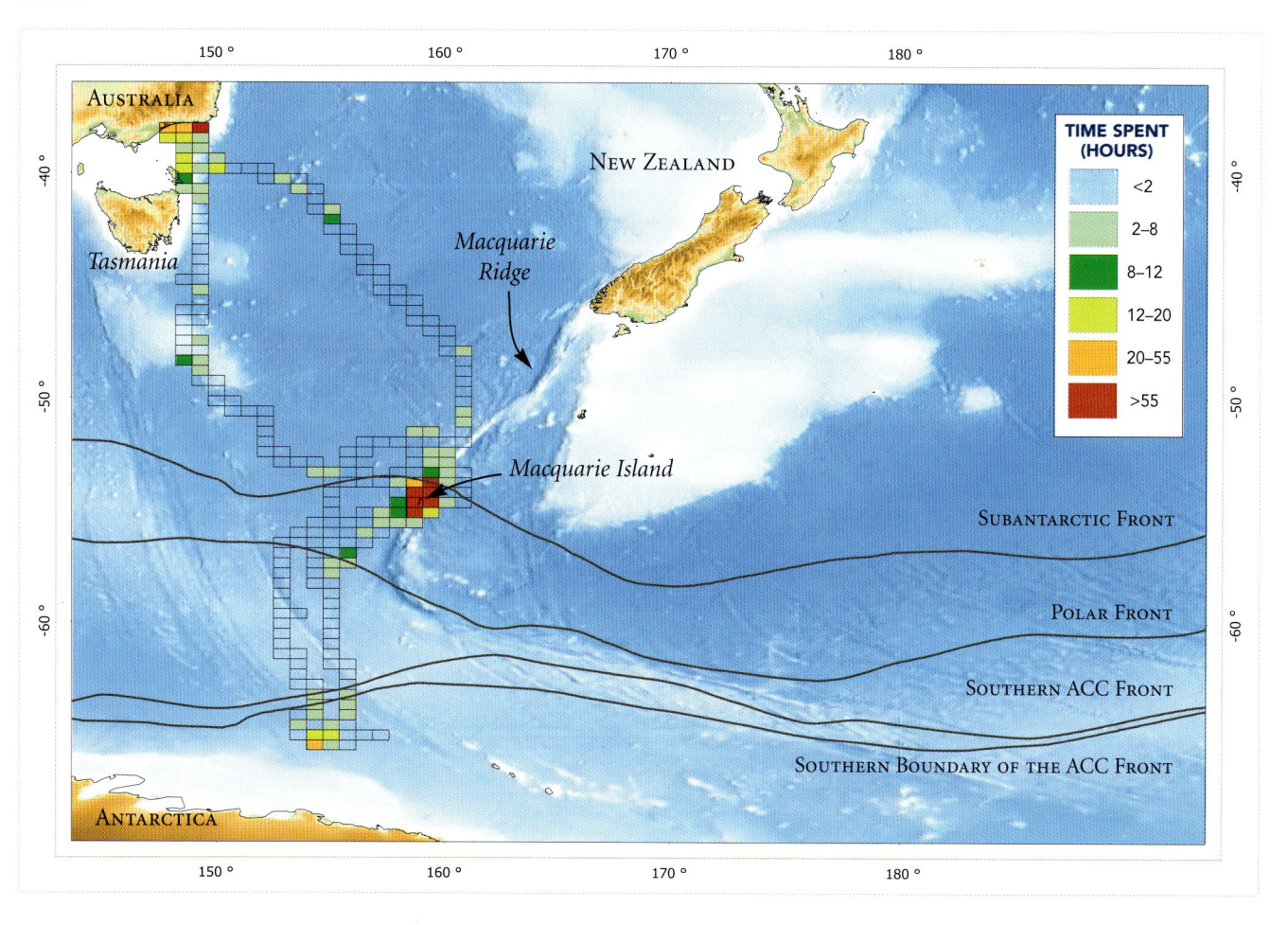

ABOVE: Is it a male or a female? Only an experienced observer may be able to tell them apart, by differences in bill and head size.

LEFT: Unlike many young birds that need to be fed every few hours, this Black-browed Albatross chick will be lucky to be fed every one to two days for its first three to four months until it fledges.

THE ALBATROSSES

Approximately 60–90 pairs of Grey-headed Albatrosses breed on Macquarie Island each year. This bird brooding its recently hatched chick stretches its wings during the long wait for the return of its partner.

Grey-headed Albatross

Thalassarche chrysostoma

Breeding Sites
South Georgia, Chilean offshore islands, Iles Kerguelen, Iles Crozet, Marion Island, Prince Edward Island, Campbell Island, Macquarie Island

LIFE HISTORY

Grey-headed Albatrosses are similar in size to the Black-browed Albatross and are closely related. They differ markedly in appearance, most notably in the head and bill colour. This species, as its name suggests, has a lustrous grey sheen to its head and this, combined with the brightly coloured orange or yellow edges to its mostly black bill, make this albatross one of the more aesthetically striking species.

Grey-headed Albatrosses on Macquarie Island typically have quite a mellow temperament and have even been known to preen the hair of researchers as they allow their leg bands to be read at the nest.

Most Grey-headed Albatrosses on Macquarie Island breed on steep, tussocked slopes, building nests on or between tussock pedestals. The breeding season begins in late September to early October, with males arriving back first, followed by the females a week or so later. Nest-building follows a similar pattern to the other albatrosses: the male finds a site and begins to build, and when the female arrives they finish off the nest together.

Grey-headed Albatrosses often return to exactly the same nest and after a few years of cumulative nest-building the bowls can reach heights of 30–50 centimetres. The slopes on which these species nest are quite dynamic and sometimes the same nest cannot be rebuilt from one year to the next as it has literally slipped off the slopes or been destroyed by small landslips that occasionally occur. Landslips have become more prevalent over the last few years as rabbit-grazing damage has increased on the southern slopes of Petrel Peak, and this has affected the

The Grey-headed Albatross has an enourmous 2-metre wing span, shown here fully stretched as the bird takes off from Macquarie Island.

A Grey-headed Albatross broods a large chick on Petrel Peak, Macquarie Island. The chick is now too large to be comfortably brooded and will soon be left unattended by its parents.

Culminated Albatross (*Diomedea culminate*), 1884. Watercolour by H. C. Richter in Gould's *The Birds of Australia*. National Library of Australia.

areas on which Grey-headed Albatrosses traditionally nest.

Eggs are laid over a relatively short period— from early to late October. The female usually does a short first shift of one or two days after which the male changes over and does a relatively long shift of up to two weeks (occasionally longer). The average shift length during early incubation is four to six days on Macquarie Island and lengthens to over seven days in later incubation.[3]

Grey-headed Albatrosses tend to forage considerably further away from Macquarie Island than their close relative the Black-browed Albatrosses, so their shift lengths are considerably longer. After the egg has hatched however, Grey-headed Albatrosses tend to forage closer to the island, and as a result the shift length decreases to two to three days.[3] The nutritional requirements of the young chick are such that it needs to be fed at this sort of frequency to survive, although some chicks can last several days between feeds, particularly as they get older.

Grey-headed Albatrosses have shown uneven breeding success on Macquarie Island, particularly during the incubation stage. Throughout the mid to late 1990s the average breeding success during incubation ranged from 55 to 80 per cent and averaged around 70 per cent.[3] Incubations fail for several reasons. One of the most common is probably abandonment of the egg by one of the pair. This usually occurs if the resident bird gets too hungry and cannot wait for its partner any longer. The eggs that survive generally hatch in December and chicks are left alone two to four weeks after hatching.

Chicks usually fledge in April and May, a little later than the Black-browed Albatross. This is probably linked to the different foraging strategies of the two species and, as the

THE ALBATROSSES 121

Set against the backdrop of a restless sea on Macquarie Island's rugged west coast, this Grey-headed Albatross rests quietly in among the tussock grasslands of Petrel Peak's western slopes.

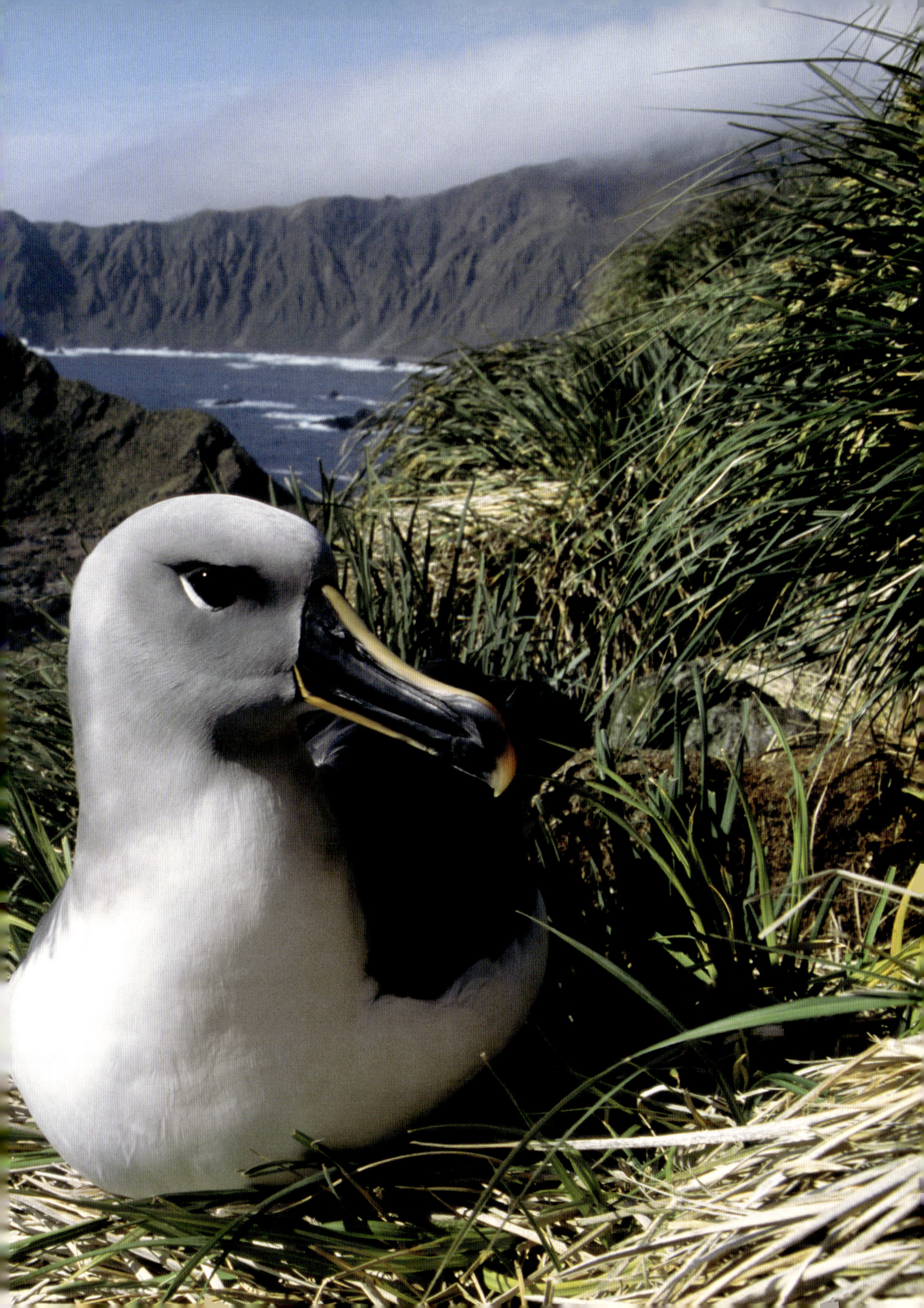

The wild and rugged coastline of Macquarie Island forms a backdrop for this lone Grey-headed adult sitting on an empty nest.

Grey-headed Albatrosses do not feed their chicks as frequently as the Black-browed Albatrosses, their chicks do not grow quite as quickly and so they fledge later in the year. Chick success is also variable with this species but it is usually considerably higher than hatching success.

Almost nothing is known of the movements of Grey-headed Albatross fledglings from Macquarie Island. Of all the albatrosses breeding on Macquarie Island, fledglings from this species spend the longest time at sea and many are not resighted again until they are over ten years old. Occasionally birds around five years old are seen but most juveniles spend many years at sea before returning to land. Once they do return to land, Grey-headed Albatrosses interact with one another, learning how to form pair bonds and build nests.

As with many other albatross species, how regularly Grey-headed Albatrosses breed is closely related to the fate of their last breeding attempt. Over 50 per cent of Grey-headed Albatrosses show a biennial breeding pattern when successful;[24] that is, once they had successfully fledged a chick they left the island and did not return for 14–18 months. Some

An adult broods its chick while an older unattended chick looks on. Older chicks sometimes have to survive alone for several days between feeds.

ABOVE:
This Grey-headed Albatross is coming in to land watched by a Black-browed and Grey-headed Albatross. The tussocked slopes of Petrel Peak, Macquarie Island, provide plenty of protected sites for nests.

pairs defer breeding even longer after successfully breeding. In fact around 20 per cent of successful breeders delay breeding for more than two years. Because of the relatively long time it takes to fledge a chick, this species will rarely breed successfully then breed again the following year. However, when pairs fail in the egg or early chick stage, most (around 60 per cent) return the following year to breed.[24] Grey-headed Albatrosses tend to form strong pair bonds and 'divorce' is rare, with most partner changes caused by the death of a mate.

The annual adult survival of Grey-headed Albatrosses is relatively high, averaging approximately 97 per cent between the mid-1970s and mid-1990s.[23] In contrast, juvenile survival is low, with an average survival rate of 34 per cent in their first years of life at sea.

Juveniles may suffer greater mortality as they learn to forage and find food successfully, and may be more at risk from interactions with fisheries due to the areas in which they forage.

POPULATION TRENDS

THE GREY-HEADED ALBATROSS population breeding on Macquarie Island appears to have remained relatively stable since the 1970s, when data on breeding numbers was first collected. It is often difficult to accurately ascertain breeding numbers of biennially breeding species such as the Grey-headed Albatross through annual counts because of the delaying of breeding by some pairs. Although breeding numbers appear variable, there is no strong evidence of a decline or increase in the number of pairs in the breeding population since the late 1970s.[23]

The flight speed of this Grey-headed Albatross may approach 90 kilometres per hour, as recorded by satellite-tracking.

A Grey-headed Albatross delicately attends to a small chick. Juvenile survival is low, with an average survival rate of 34 per cent in their first years of life at sea.

LEFT:
Grey-headed albatrosses nesting in the sun on West Rock, a small rocky outcrop at the base of the western slopes of Petrel Peak, Macquarie Island.

BELOW:
The soft grey on the head of the Grey-headed Albatross extends down the neck onto the back.

ABOVE:
A pair of Grey-headed Albatrosses takes a break from nest-building. They often return to the same nest so after a few years nests can reach heights of up to 50 centimetres.

THE ALBATROSSES

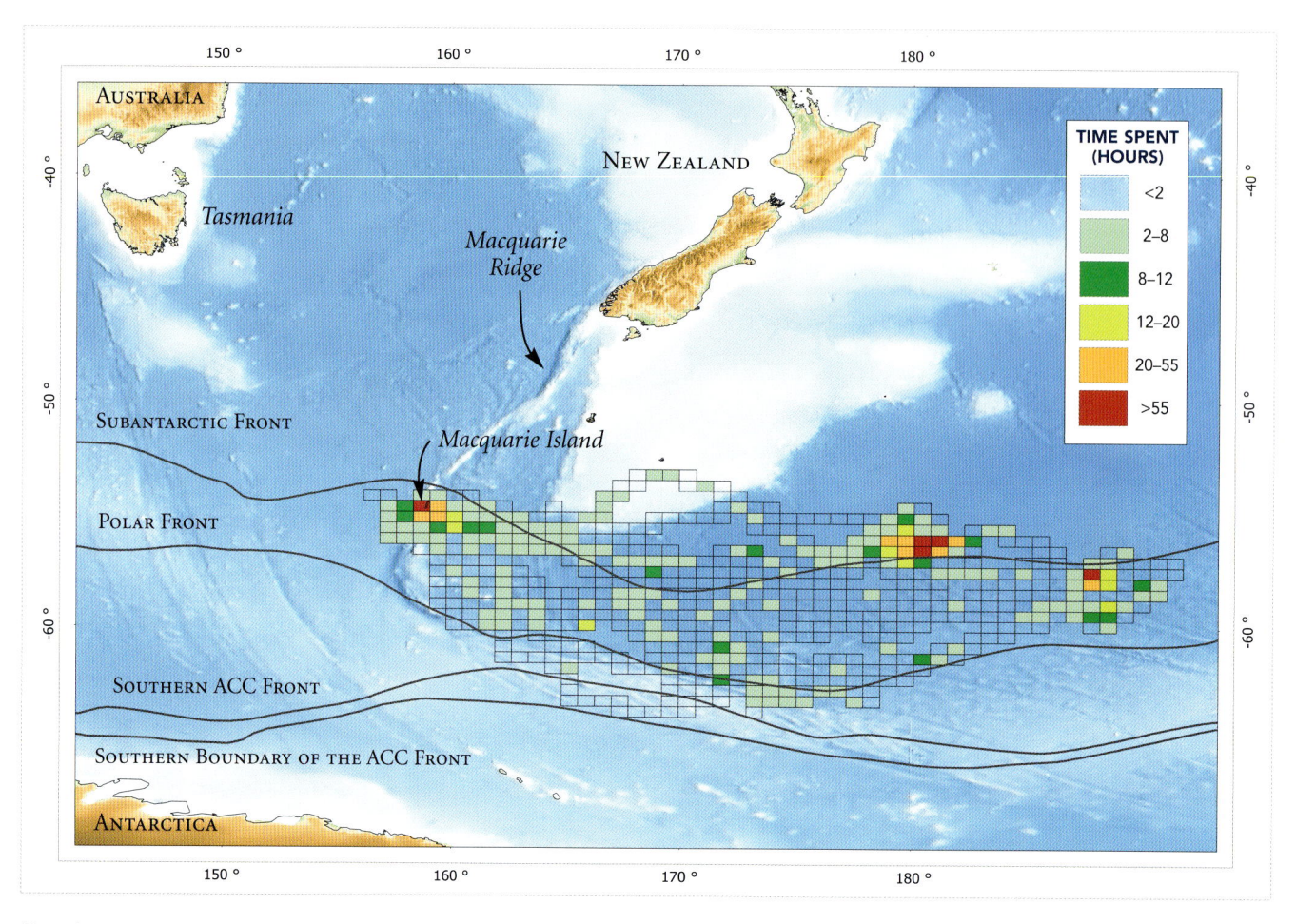

Foraging areas of Grey-headed Albatrosses breeding on Macquarie Island.[29,34]

FEEDING HABITS

SEVEN BREEDING GREY-HEADED Albatrosses were satellite-tracked between 1998 and 2001 and these birds showed quite a different foraging pattern to the Black-browed Albatrosses tracked at the same time. During incubation most of the tracked Grey-headed Albatrosses foraged over two main areas some 1500–2100 kilometres to the east and east–south-east of Macquarie Island (see map above).[29] These travelling distances are similar to those described by South African researchers, who found that the average distance travelled by this species in a single long-distance voyage breeding on Marion Island was 2182 kilometres.[30] Accurate satellite-tracking from Macquarie Island showed that birds moved at speeds approaching 90 kilometres per hour while travelling to distant foraging locations, and travelling almost invariably began soon after changeover. The speed of travel on the return journey was generally slower, and birds followed a less direct route to Macquarie Island, suggesting that they may have been prospecting for, and foraging on, more sporadically distributed or patchy resources. Most foraging during incubation was conducted in the vicinity of the Subantarctic Front in the Polar Frontal Zone and, to a lesser extent, the Polar Front.

Grey-headed Albatrosses appear to target different oceanographic features from Black-browed Albatrosses, which is likely to be related to the different prey items of the two species. Underwater topography appeared to play a much less important role in the areas targeted by this species during incubation since most foraging occurred over deep, relatively featureless waters to the east of the Campbell Plateau.[29] These waters are rich in evolving and decaying eddies (described in Chapter 2), and most foraging is concentrated at the edges of these features or at their interfaces. Again, this is very similar to the foraging strategies of Grey-headed Albatrosses on pelagic trips from Marion Island where birds like to feed around warm eddies in the Polar Frontal Zone.[30] Both warm and cold eddies were targeted by birds from Macquarie Island and were also often associated with high gradients in sea-surface temperatures.

The brightly coloured stripe on the black bill of the Grey-headed Albatross makes it a handsome bird. This one has picked out a potential nest site on the southern slopes of Petrel Peak.

The Light-mantled Sooty Albatross is smaller and darker in colour than the other Macquarie Island species.

Light-mantled Sooty Albatross

Phoebatria palpebrata

Breeding Sites
South Georgia, Iles Kerguelen, Iles Crozet, Marion Island, Prince Edward Island, Auckland Islands, Antipodes Islands, Campbell Island, Macquarie Island, Heard Island.

LIFE HISTORY

Light-mantled Sooty Albatrosses are the smallest of the Macquarie Island albatross species, with adults weighing 2.5–4 kilograms. Their appearance differs from the other albatrosses in several ways. The most obvious difference is their plumage, which is dark all over except for a lighter shade of grey on their back and abdomen (from which they get their name). Their close relative, the Sooty Albatross, is the only albatross with similar pervasive dark plumage but it is darker still, without the lightening of the grey colouring on the back.

Although the male Light-mantled Sooty Albatross tends to return to the nest site first and prepares it for the female, as do other albatross species, this species has a number of distinguishing behavioural features. First, it has a very distinctive two-toned call that is used to attract other birds. The cry is usually heard before the first birds are seen and its sound heralds the end of winter for many of the expeditioners living on the island. It is accompanied by an equally distinctive rapid, up-and-down movement of the head, which it is also able to perform in mid-air while flying. The 'sooties'—as they are often known—also have a complex courtship flight: two, occasionally three birds will fly in complete synchrony in intricate flight patterns. This species has a longer tail than most other albatrosses, allowing it to manoeuvre more easily in the air. Once new pair bonds are formed or old pair bonds are re-established, pairs settle on a nest site and get down to preparing the nest for the egg.

The nests on the coastal slopes are built in and around various types of vegetation, from tussock grasses to ferns. Birds also nest in some of the most inaccessible nesting sites, such as on small ledges or rock faces that are barely large enough to accommodate the nest. Like the other albatrosses, 'sooties' construct their nests out

A Light-mantled Sooty Albatross calls to a flying bird on the slopes north of Hurd Point, Macquarie Island. The distinctive two-toned call of the Light-mantled Sooty Albatross is unique to the species and is used to attract potential mates.

of mud or vegetation and often build them onto existing nests from previous years. The first eggs are laid during November, and following a typical short shift on the nest during which the female lays her egg, she is relieved by the male.

The shifts are of variable length during incubation and birds spend on average 10 days at sea each time.[3] The periods away can be much longer, though, and some birds have been observed sitting on the nest for 25 days or more before their partner returns to change over. With an incubation period of about 66 days, the eggs hatch around the end of December or the beginning of January. This somewhat shorter-than-usual incubation period is related to the smaller size of this species. On average, about 50 per cent of all eggs survive, but breeding success between sites and between years is extremely variable.[3]

Once hatched, the chick is attended constantly by one or other of the parents and they shorten their shifts to around two days. This brood-guard period lasts between 15 and 25 days, after which the small chicks are left to fend for themselves. A Light-mantled Sooty Albatross chick that has been left alone for the first time can look very small in the nest, but they are by far the most tenacious of all the albatross chicks. Nevertheless, they are more prone to skua predation than other chicks on Macquarie Island, and occasionally several skuas will work together to make a small chick fall out of its nest. Once an albatross chick has left the safety of its nest, it is extremely unlikely to survive. Even if it manages to stay alive, the parents will typically only feed a chick that is sitting on its nest or within a metre or so of it.

Once the chick is unattended, parents undertake a combination of long and short foraging trips. The long foraging trips can last up to 20 days. If both parents are on long trips the chick may not get fed for a week or more.

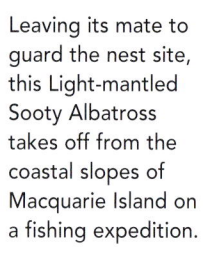

Leaving its mate to guard the nest site, this Light-mantled Sooty Albatross takes off from the coastal slopes of Macquarie Island on a fishing expedition.

TOP
Sooty Albatross (*Diomedea fuliginosa* Gmel.), 1848. Watercolour by H. C. Richter in Gould's *The Birds of Australia*. National Library of Australia.

RIGHT:
'Sooties' build their nests from mud or vegetation, often refurbishing second-hand nests from previous seasons.

THE ALBATROSSES 135

A pair of non-breeding Light-mantled Sooty Albatrosses on Macquarie Island. Having raised a chick in the previous year, this species typically takes a year off before attempting to breed again.

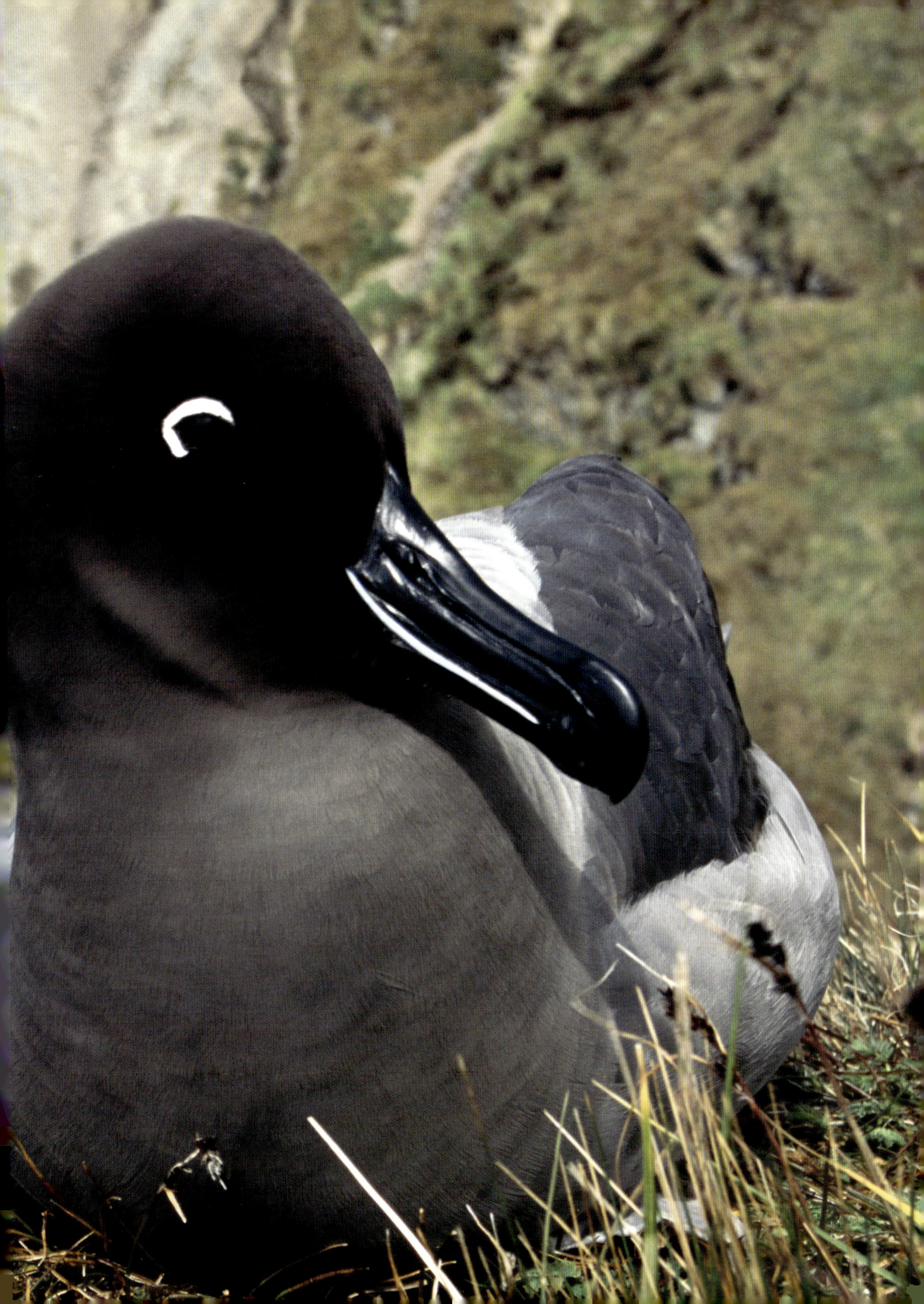

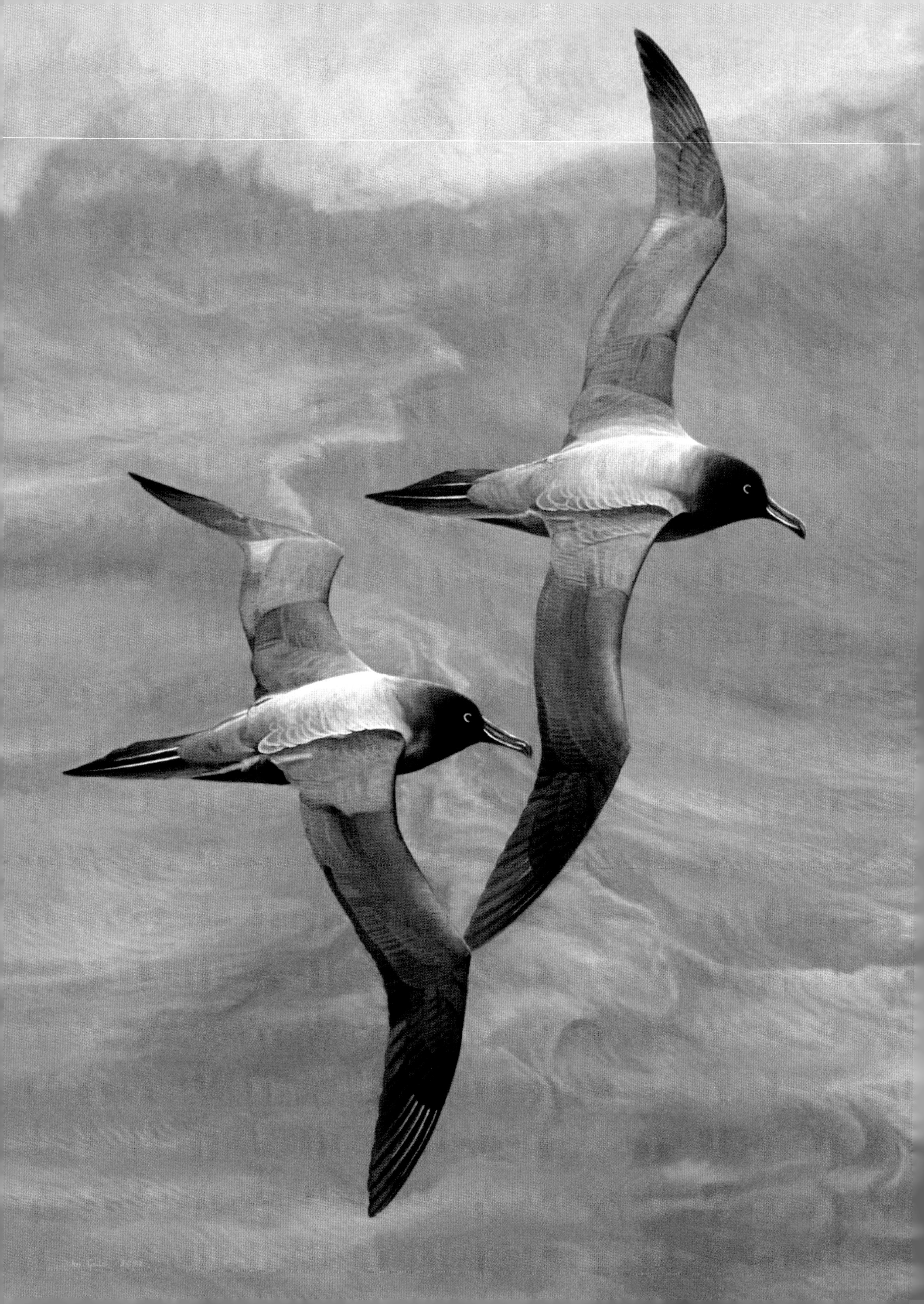

The amount of food that a parent brings back corresponds to the amount of time it has spent at sea, so meals may range in size from just a few hundred grams to over a kilogram. The chicks are fed 30–40 kilograms of food over the provisioning period, which results in a weight gain of 1–2 kilograms.[31] Hatchlings weigh just a few hundred grams but by the time the 'sooty' chick is fledged it usually weighs just over 3 kilograms. They reach their peak weight after 60–100 days; thereafter they undergo a gradual weight loss while they metabolise their large reserves of fat into muscle. The chicks are generally fed squid, small fish or krill. Small squid beaks are often scattered around the nests as the chicks regurgitate them after they have digested the soft parts.

Most breeding failures on Macquarie Island occur soon after egg-laying or around the time when the egg is due to hatch and through the brood-guard period. Factors such as prevailing wind, location of nests and vulnerability to predators and ticks may all have some impact on breeding success. So, too, may the fact that birds breeding at different sites may forage in different areas and this could also be linked to changes in attendance patterns and subsequently nest desertion, which appears to be more frequent at sites of generally lower breeding success. Variations in breeding success over time on Macquarie Island and the differences between the different locations are complex. It may well be that the way in which birds are feeding, and particularly the availability of prey, are major factors. Breeding success also depends upon such things as the experience of the breeders, partner fidelity, chick provisioning regimes and individual fertility.

The chicks fledge after about 140 days,[31] and in the weeks prior to fledging they spend much of their time exercising their wings and building up their flight muscles. As fledging approaches, the chicks begin to tentatively step off the nests and begin exercising more frequently until finally a particularly vigorous set of flaps or a big gust of wind lifts them into the air. The fledglings quickly adapt to life in the air and will not return to land for another four to eight years. Little information on the survival rates of either juveniles or adults is available for Macquarie Island, so how many young birds survive during the first few years of their life remains unknown.

The breeding frequency patterns of Light-mantled Sooty Albatrosses are superficially similar to those of the Grey-headed and Wandering Albatrosses, with a large proportion of the population breeding on a biennial basis. Frequency depends on the result of the previous breeding attempt. For example, those pairs that have successfully raised and fledged a chick will almost always take at least one year off before attempting to breed again. Around 20 per cent of the population will defer breeding for more than a year and some pairs have as many as four years off before breeding again.[24] Pairs that have not fledged a chick in the previous year are much more likely to attempt to breed the following year; in fact over half of all unsuccessful breeders will attempt to breed the following year. Breeding deferral is still observed in unsuccessful breeders with up to one-quarter of all birds having a year off after an unsuccessful breeding attempt.[24] Many pairs defer breeding for longer and some pairs have been recorded taking up to four years off between breeding attempts.

OPPOSITE: Light-mantled Sooty Albatrosses in courtship flight, 2005. Oil on Board. John Gale. Private collection.

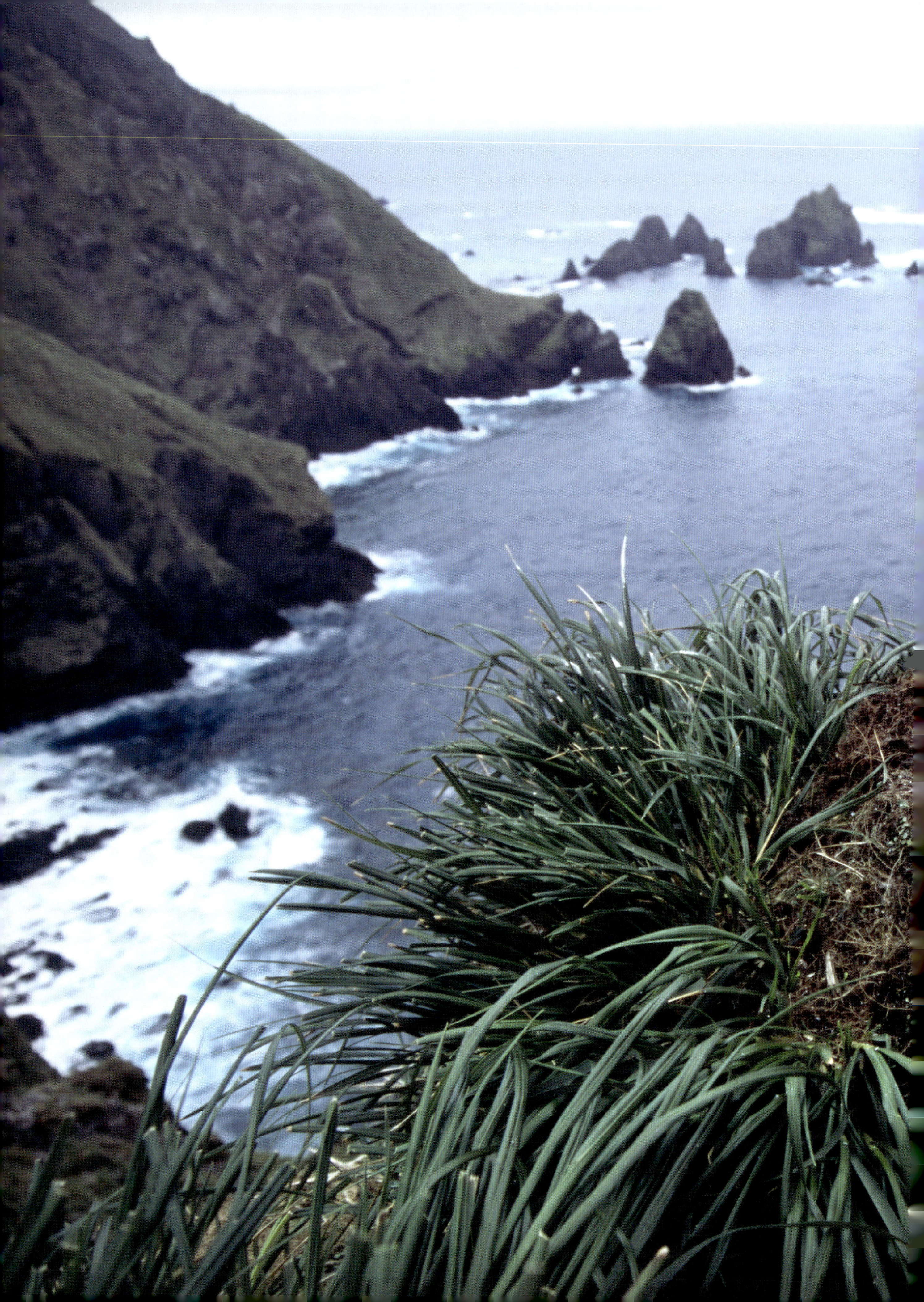

This Light-mantled Sooty Albatross chick is only a few weeks old. Left unattended at such a young age it is vulnerable to the unwanted attentions of the carnivorous skuas that regularly glide over colonies of young birds.

Light-mantled Sooty Albatrosses have a longer tail that allows for more manoeuvrability in the air than other albatrosses.

POPULATION TRENDS

VERY LITTLE IS KNOWN ABOUT the population trends of Light-mantled Sooty Albatrosses on Macquarie Island because there are comparatively high numbers of breeding birds and it is not possible to easily count all the nests each year. The monitoring that has been conducted so far suggests that numbers are stable but, without good data on total population numbers over several years, it is difficult to make an accurate assessment of what is happening to the population.

FEEDING HABITS

LIGHT-MANTLED SOOTY ALBATROSSES were first tracked by researchers on Macquarie Island in 1993.[32] Five birds were tracked during the incubation stage and they travelled up to 2000 kilometres in a single trip on their search for food. The birds left their nests and generally travelled quickly south or south-west until they met the colder waters that marked the Polar Front, or even further south to those fronts that mark the southern boundary of the Antarctic Circumpolar Current. They tended to travel along these frontal zones, probably feeding on prey congregating at the interfaces between the cooler and warmer waters. After continuing along the frontal zones for hundreds of kilometres they mostly undertook a long, looping flight path back to Macquarie Island.[32]

The shorter shifts that are observed during the brood-guard stage indicate that the parents are likely to be feeding much closer to Macquarie Island. In fact this pattern is observed in every species of albatross breeding on Macquarie Island and it is likely that the nutritional demands of the small chick are so great that the parents are forced to find food closer to home. Nevertheless, some birds do undertake longer flights while the chick is still very young, suggesting that occasionally they cannot find adequate food close to Macquarie Island.

Once the chick has hatched, the parents undertake a cyclical pattern of long and short trips. The cycle usually consists of one long trip—usually around six days but up to 20—followed by several short trips.[31] Shorter trips tend to be more energy expensive for the parents and the long trips allow the parents to rebuild reserves that they have depleted during the breeding season.

The nutritional requirements of the chick create an internal conflict within the parents as they try to deliver enough food to keep the chick alive without draining their own resources too much.[33] The cyclic strategy of long and short trips probably allows the parents to maintain their own condition while keeping the chick well fed and alive.

It's easy to make out the lighter shade of grey on the back and abdomen that help distinguish the Light-mantled Sooty Albatross from other species.

This Light-mantled Sooty Albatross chick, only weeks from fledging, has been moved onto an automatic weighing nest. Researchers have simply removed the original nest and replaced it with the new one. The birds blithely continue their lives as if nothing has happened, while researchers are able to gather valuable data.

CONSERVATION

MOST ALBATROSS POPULATIONS HAVE EXPERIENCED declines over the last two to three decades, but through the combined efforts of many people some of the threats to the birds have diminished and some populations have slowly started to recover.

Hundreds of people around the world have dedicated much of their life to working on the conservation of albatrosses and other species threatened by human activities.

Long-term comprehensive studies on subantarctic populations of albatrosses have been conducted on French islands in the southern Indian Ocean and by British researchers on South Georgia in the Atlantic Ocean since the 1970s. The information and publications that have been produced by these studies have not only illustrated the impact of long-line fisheries on albatross populations but also contributed greatly to our knowledge of the breeding biology and foraging ecology of albatrosses in general.[1,2]

More recently, subantarctic studies have documented population dynamics, breeding biology and foraging ecology of albatrosses on New Zealand subantarctic islands[3,4,5] and on South Africa's Marion Island in the Indian Ocean.[6,7] Studies on critically endangered species, such as the Amsterdam Albatross (*Diomedea amsterdamensis*), have also highlighted the importance of monitoring in identifying populations that are at risk.[8] These studies by no means encapsulate all research done on the conservation of albatrosses globally, but they do indicate the breadth of knowledge and understanding that has accumulated, particularly in the subantarctic. Nevertheless, there are still significant gaps, particularly at the population level. Only at a population, rather than a species, level can the global conservation status of any species be assessed. In addition to the population monitoring studies, scientists and fisheries observers have conducted much valuable research on long-line fisheries around the world. This information is extremely useful in developing techniques to both quantify and minimise the by-catch of seabirds.

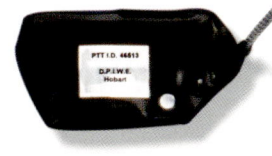

The following sections describe some of the research that has been conducted in Australia, focussing specifically on Australian researchers and the albatrosses breeding on the Tasmanian islands and on Macquarie Island. This research has contributed significantly to the global pool of data. Complementing and adding to the global data set in this way is critical in assisting global conservation bodies—such as the International Union for the Conservation of Nature (IUCN), BirdLife International and the Agreement on the Conservation of Albatrosses and Petrels (ACAP)—to focus conservation efforts on the most relevant issues.

This female Wandering Albatross has been fitted with a satellite tracker (pictured far left) on its back. These trackers gather valuable information about where the birds feed.

TASMANIAN ISLANDS

NATURALISTS BEGAN TO VISIT THE islands on which the albatrosses breed in the 1800s and, although these early visits may not have been based on conservation as we think of it today, they started a process that culminated in the conservation efforts of later years. On Albatross Island, these naturalists identified the albatrosses as Shy Albatrosses (or White-capped Albatrosses, as they were then known) and collected information on breeding numbers and basic breeding biology. Due to their remote location and inaccessibility, the albatrosses breeding on the Mewstone and Pedra Branca were not identified as Shy Albatrosses until much later.

Researchers visited Albatross Island sporadically following these first visits; however, landings were still difficult and visitors rare. In 1960, two researchers visited the island with the aim of studying the albatrosses and other wildlife breeding there.[9] They made population estimates of the albatrosses and banded some of the nestlings. They also documented other breeding birds, plants and skinks. Another visit was made in 1973 by a team composed of an ABC film crew and ornithologists.[10] These researchers also made estimates of breeding numbers, breeding success, chick provisioning and the timing of the breeding cycle. They also began to document some of the breeding behaviour of the Shy Albatrosses.

Nigel Brothers, from the Tasmanian Parks and Wildlife Service, initiated a long-term monitoring program in 1980 that continues today, and his efforts have contributed greatly to the conservation of this species. He and co-workers established study plots and banded chicks each year. This work has been continued and maintained in more recent times by Rosemary Gales. The establishment and maintenance of this program has allowed the population demography of the Shy Albatrosses to be investigated. Because of their characteristic life histories, it is impossible to assess population trends over short periods of time, so long-term studies are the only way to make accurate assessments. The establishment of these early study plots allowed this to happen with the Shy Albatrosses.

In 1992 the Australian Conservation Agency commissioned Rosemary Gales to write a landmark review of populations of albatrosses around the world, the threats they face and possible ways of reducing these threats.[11] This review showed that, while little was known about some of the albatross populations, many of those for which there was information were declining.

A couple of weeks before a chick leaves the island, researchers attach a band to its leg (pictured far right) to help monitor albatross populations.

Fisheries were identified as the key threatening process and Nigel Brothers was one of the first to quantify the extent and magnitude of the problem.[12] He and other fisheries observers around the globe gathered information on bird by-catch rates and, when combined with the population studies that were also being undertaken around the world, it became clear that hundreds of thousands of albatrosses had been caught and killed on long-line fishing boats since this practice had begun in the 1960s.

In 1993 studies were initiated into how Shy Albatrosses were feeding and satellite trackers were attached for the first time to breeding adults.[13] In 1999 Rosemary Gales, April Hedd and Nigel Brothers initiated a more focussed study into the Shy Albatrosses at all three of their breeding colonies. In this study they satellite-tracked breeding birds and in so doing obtained valuable information about how the birds were feeding, breeding and raising their chicks. The data they collected showed conclusively that Shy Albatrosses were foraging in areas that significantly overlapped with the Japanese long-line fishery in Australian waters and were therefore at serious risk from long-line fisheries.[13,14]

Research continued into the by-catch levels in Australian waters and showed that the Japanese tuna fleet were impacting upon 15 different seabird species and catching thousands of birds each year.[15] Working closely with fishers, researchers have tried to reduce the by-catch of seabirds. They have introduced measures such as reduction of offal discharge, streamer lines that scare the birds and weighted lines that reduce the amount of time the bait stays on the surface of the water. Graham Robertson, an Australian researcher, has been particularly proactive in working with fishers to increase the use of weighted lines and help reduce seabird by-catch.

Aleks Terauds gets up close with an aggressive Black-browed Albatross as he tries to check the band around its leg. Photo: Tore Pedersen.

Tracking of individual birds will help with the conservation of the species. This Shy Albatross fledgling has had a satellite tracker attached to its back and plastic band around its leg.

In addition to those already mentioned, Barry Baker from the federal Department of Environment and Heritage has supported research in both Tasmania and on Macquarie Island, as well as being active in the development of the Agreement on the Conservation of Albatrosses and Petrels. Fisheries observers also play a critical role in gathering vital information and their contribution has also been invaluable. More recently, Sheryl Hamilton and Rachael Alderman have contributed significantly to the ongoing research into Shy Albatrosses, and Mike Double and Cathryn Abbott have conducted important genetic studies to clarify the taxonomy of albatrosses.

The listing of long-line fishing as a key threatening process under the Australian Environmental Protection and Biodiversity Conservation Act was a significant step towards albatross conservation in Australia. Once these fisheries were listed as a threatening process a Threat Abatement Plan was commissioned and a number of compulsory regulations were eventually adopted in Australian waters. Although there are still some problems with the by-catch of smaller petrels in long-line fisheries, as a result of these regulations the by-catch of albatrosses has been reduced in Australian waters.

More recently, conservation efforts in Tasmania have taken a slightly different turn, with the advent of the Big Bird Race. Funding was obtained from Ladbrokes, a high-profile betting firm from the United Kingdom, to sponsor a 'race' of albatrosses. In March 2004, 18 Shy Albatross chicks were fitted with transmitters by researchers from the Department of Primary Industry, Water and Environment and people could bet on the first bird to cross the finish line, arbitrarily set at the Cape of Good Hope. Ladbrokes agreed that after covering costs, any money that was made out of the betting would be used for research into albatross conservation.

Very little was known about the movements of Shy Albatross chicks, with only a few recoveries of banded birds suggesting that they were travelling across the Indian Ocean in the first few months of their time at sea.[16] This is considerably further than the breeding or non-breeding adults were likely to travel, so the chicks were likely to be in much greater danger from fisheries in the unregulated waters of the high seas.

Satellite transmitters weighing 45 grams or less were attached to the fledglings a week or two before they left the nest and programmed so that they would transmit for several hours each day. Although the transmitters only had a battery life of approximately 30 days, programming

Tasmanian marine biologist, Rachael Alderman carefully weighs a Shy Albatross fledgling before attaching a satellite tracker.

Wade Fairley shooting documentary footage associated with the Big Bird Race in which people lay bets on which of 18 birds fitted with satellite trackers would reach the Cape of Good Hope first. Proceeds from the race were to be used to fund research into albatross conservation.

them in this manner allowed several months of information to be gathered. Further funding was obtained from Ladbrokes and 20 more transmitters were attached in April 2005. Once the data from these transmitters has been processed, important information on the foraging areas of these birds will be obtained and this will assist considerably in the conservation of this species.

MACQUARIE ISLAND

LONG-TERM SCIENTIFIC INVESTIGATIONS of Macquarie Island began in 1911, with the Australian Antarctic Expedition 1911–14, led by Douglas Mawson, establishing a major base on the island. The purpose of this base was to set up a combined wireless relay station and meteorological observatory.[17] Ever since 1949, Australian National Antarctic Research Expeditions (ANARE) have taken place annually.

Research into the albatrosses began in the 1950s. Breeding censuses were done and data on basic breeding biology was collected. The banding of Wandering Albatross chicks began, too, around this time and continues to this day, making it one of the longest albatross monitoring programs. Banding individual birds is important because it allows individuals to be followed over time, thereby providing vital information about population trends and survival.

ANARE researchers continued to study the albatrosses and petrels throughout the 1970s and 1980s and, as banding studies became more focussed, more information came to light about how the birds were breeding and how populations were faring. Gavin Johnstone, Knowles Kerry, Bob Tomkins and many others contributed greatly to research on the Wandering and Light-mantled Sooty Albatrosses. In the early 1990s that information allowed the trends of the Wandering Albatross population to be comprehensively described and the impact that long-line fishing vessels were having quantified.[18]

Geof Copson studied the Black-browed and Grey-headed Albatross populations at Petrel Peak through the late 1970s and early 1980s and was one of the first to publish his data on the population trends of these species on Macquarie Island.[19] In 1993 Rosemary Gales and Nigel Brothers initiated a long-term population study into the albatrosses and petrels breeding on Macquarie Island.

Fieldwork on Macquarie Island began in 1994. A team of four people, led by Jenny Scott, set up the project on the ground in a season that lasted from September 1994 to May 1995. Annual trips have been conducted ever since, with two people working on the project each summer season. Field seasons for individual workers have typically lasted from three to nine months, with occasional longer stints of up to 17 months. Data on population trends, breeding biology and foraging ecology were the main types of information collected. The primary aim of the project was to ascertain the conservation status of the albatross and giant petrel populations breeding there.

I joined the program in the third year (1996) and have been involved in six summer seasons since that time, spending approximately four years in total on Macquarie Island. In 1998 I enrolled in a doctoral degree at the University of Tasmania and, in conjunction with Rosemary Gales and Nigel Brothers, conducted further research into the albatrosses on Macquarie Island.

The main aims of the doctoral study formed part of the broader aims of the long-term study and were to: 1) identify and quantify trends in breeding numbers and survivorship in the vulnerable albatross populations on Macquarie Island and investigate factors that may be responsible for these trends; 2) describe aspects of the breeding ecology of albatrosses on Macquarie Island that could influence breeding numbers and breeding success; 3) investigate the foraging ecology of breeding Black-browed and Grey-headed Albatrosses; 4) describe the chick provisioning regime of Light-mantled Sooty Albatrosses on Macquarie Island; and 5) use the above data to comprehensively describe the current population status of albatrosses on Macquarie Island and identify the threats that may impact upon these populations in the future.

When comprehensively assessing the conservation status of any population, a substantial amount of baseline data from long-term studies on breeding numbers and trends is needed.[20] To assist in the investigation of the Macquarie Island Wandering, Black-browed and Grey-headed Albatross population trends, old biological logbooks were used to obtain historical information on breeding numbers. Data on the frequency of the breeding performance and foraging ecology are also important when assessing the conservation status of a population. Whenever possible, we investigated specific aspects of Wandering, Black-browed, Grey-headed and Light-mantled Sooty Albatrosses in order to make comparisons. However, due to the different sizes of the populations, and logistical and budgetary

Jason Hamill checks out nesting Light-mantled Sooty Albatrosses above a penguin colony on the coastal slopes of Macquarie Island.

constraints, this has not always been possible.

The study found that, while the population status of the albatrosses breeding on Macquarie Island could be described as currently stable, there is little room for complacency. There are still fewer than 20 pairs of breeding Wandering Albatrosses on Macquarie Island in total and this population is precariously balanced. Any extra mortality could easily send this population into decline and illegal fisheries still pose a huge threat to albatrosses on the high seas. Recent agreements such as the Agreement on the Conservation of Albatrosses and Petrels (ACAP) provide hope that conservation of albatrosses can be managed on a more global scale. ACAP has a number of potential benefits. One of the main advantages is that it can help researchers, organisations and governments understand the status, susceptibility and threats to albatrosses and petrels. ACAP will provide a framework for collecting and analysing data on these birds that can potentially be used by all countries conducting research on albatrosses.[21] Probably one of the most important things that ACAP can do is establish benchmark levels for the by-catch of seabirds and persuade countries to sign up to an agreement to maintain those levels throughout their fisheries. If these levels are then exceeded, it would evoke urgent action by all parties to the agreement.[21]

This tent houses the data logger and batteries used in conjunction with the automatic weighing nest on slopes north of Hurd Point, Macquarie Island. In the background are the wind generators that power the equipment.

ABOVE:
Aleks Terauds retreats after attaching a satellite transmitter to a female Wandering Albatross. Some birds have been known to groom researchers as they attach bands or transmitters.

LEFT:
Light-mantled Sooty Albatross chick being fed by parent on automatic weighing nest.

CONSERVATION 153

Tents are erected on this ledge of Pedra Branca to allow researchers to live among the birds with minimal disturbance to their nests. Living conditions are harsh: prevailing winds are strong and the tents are surrounded by the noise of the birds and the intense smell of guano.

Living and Working on the Islands

People have been visiting the islands on which the albatrosses breed around Tasmania and on Macquarie Island for many years, but more focussed research efforts into albatross biology and ecology have only been going on for the last few decades.

Visits to the islands require careful planning and logistical efforts. All the islands are in relatively remote areas, access is not easy and attempts to get to them are very weather dependent. The following is a brief description of what is involved in getting to the three islands around Tasmania and the conditions experienced while working and living on them.

ALBATROSS ISLAND

Albatross Island is accessible only by boat and requires good weather to land. Over the last 20 years, transport has been largely provided by Neil Smith in his ketch, the *Wild Wind*. Transport has also been provided in more recent years by the Tasmanian Marine Police, thanks largely to the efforts of Maurie Fowler. The two modes of transport could really not be more different. It takes one to two hours to reach Albatross Island in the police sharkcat and six to eight hours in the *Wild Wind*, but both are enjoyable in their own way. Given a choice, most regular visitors would choose the *Wild Wind* as, although it is slower, the view is never boring and there is always something to see.

On arrival at Albatross Island the landing is usually made via a dinghy that is expertly rowed ashore by Neil, or in the case of the police sharkcat, nosed up close to a steep-sided rock and the gear passed ashore. Enough food, water and camping gear are taken ashore for the duration of the stay. Trips generally last for three to seven days, although in the past they have been considerably longer. Researchers usually

Getting ashore on the Mewstone is no easy task. Researchers are dropped off by the Marine Police in the *Swift*, a powerful rubber duckie supported by the *Vigilant*, seen in the background.

camp in the main cave, which they share with hundreds of prions and penguins. The noise at night has to be heard to be believed and it often takes a night or two to get used to the din.

Stays on the island are busy, the days filled with monitoring of nests, reading bands or making other nest observations. During March many of the chicks on the island are banded and this takes three to four researchers several days to complete.

PEDRA BRANCA

PEDRA BRANCA IS A UNIQUE island on which to live and work. Situated on the continental margins of the Southern Ocean, it has only been visited regularly by researchers since the 1970s, when monitoring studies began on the Pedra Branca Skink. Researchers have visited the island regularly since this time and trips have been made in most years. Visiting this island is very weather dependent and a good weather window is required before any attempt is made to land and stay ashore. As for Albatross Island, the *Wild Wind* skippered by Neil Smith has provided most transport to the island in the past and in more recent years the Tasmanian Marine Police boat, the PV *Vigilant*, skippered by Scott Dunn, has also assisted in much of the transport to this and other southern islands.

Researchers generally camp in tents on the lower eastern ledge of Pedra Branca. Most visits are made after the first week in April when the gannet chicks are beginning to fledge; any earlier visits would create too much disturbance to the thousands of gannets that are nesting there. Tents are pitched on an oozy guano mud mixture on the eastern side and if the noise is the main thing that stops you sleeping on Albatross Island, the smell is many times worse on Pedra Branca. It is a little harder to get used to than the noise of the birds however, and the distinctive Pedra smell stays with most of the gear for weeks.

In fact guano is a way of life on Pedra Branca and you have to get used to having it not only on your clothes and in your gear, but also in your hair, your cups of tea and of the food you eat. There are so many birds flying about that it is almost impossible to avoid.

Trips usually last a few days and, in addition to monitoring the albatrosses and the skinks, diet samples from the gannets are collected, seal counts are made and numbers of Silver Gull nesting there are also monitored.

THE MEWSTONE

DUE TO ITS INHOSPITABLE NATURE, people do not typically stay on the Mewstone, although some researchers have stayed there for short periods of time on a couple of occasions in the past. More recently, day trips are typically made

Researchers share a cave on Albatross Island with hundreds of prions and penguins. It's noisy but at least their gear stays safe and dry.

LEFT:
Neil Smith rows back to his boat, the *Wild Wind*, after picking up gear and people from Albatross Island.

once every year or so to conduct monitoring of the albatrosses breeding there. There are a number of painted rocks on the Mewstone that mark out five 30-metre by 30-metre squares. Aerial photographs are taken of these squares each year and used to monitor both population numbers and the breeding success of the Shy Albatrosses. Visits to the Mewstone are kept to a minimum these days to minimise disturbance to this pristine island that has seen very little human presence since its discovery.

MACQUARIE ISLAND

Macquarie Island takes a bit more logistical effort to get to and typically involves the time and resources of large organisations like the Australian Antarctic Division (AAD). The AAD vessels have traditionally been the most common way of travelling to Macquarie Island, and in recent times, the RV *Aurora Australis* has most commonly been used. Every year hundreds of researchers, tradespeople, maintenance crews, Bureau of Meteorology staff and others make their way south on this vessel to one of the four ANARE bases. The Macquarie Island station is a relatively small one by Australian standards and houses a maximum of 40 people over the summer months. During the winter the numbers vary but, in recent years, there have been as few as 12 and as many as 25 people.

Over the last few years, the number of AAD vessels that travel to Macquarie Island has been reduced and currently only one resupply visit is made each year, usually in March. Therefore scientists and others wishing to get to Macquarie Island for a summer season utilise tourist ships that now visit the island each year. Currently around 500 tourists visit Macquarie Island each year and, although they do not stay on the island itself, they are able to get ashore for at least a few

BELOW:
The main ANARE station on Macquarie Island provides a relatively comfortable home to up to 40 researchers all year round, despite the harsh climate.

Living and Working on the Islands   157

RIGHT:
Researchers staying at Hurd Point field hut get to share their environment with hundreds of Royal Penguins.

BELOW:
Researchers finish their day with an exhilarating run down the scree slopes to the hut at Hurd Point.

LEFT:
Hurd Point Field Hut even has a balcony, from which researchers can enjoy the view (below left).

BELOW RIGHT:
Life inside Hurd Point field hut is simple, but comfortable.

hours, see the main station and surrounding wildlife and get a real taste of the subantarctic.

On arriving at Macquarie Island the new expeditioners are briefed at a meeting about living and working on the station. Although most people have had some form of field training in Hobart before departing, further training is usually required before venturing away from station limits into the Macquarie Island environment.

The main station consists of a number of buildings on the low-lying isthmus at the northern end of the island and expeditioners sleep in rooms of varying sizes known as dongas. It is a relatively comfortable life on the main station even though the climate can be harsh. Macquarie Island has a mean annual temperature of 4°C. It rains most days and winds above 30 knots are not uncommon. The station offers most of the comforts of home, with a small library, video rooms, a pool table and other communal areas. The main communal area is known as the mess where meals are taken and festive occasions are celebrated.

A small number of people, including the albatross researchers, live off the main station in field huts scattered around the island. There are currently five field huts on Macquarie Island—at Brothers Point, Bauer Bay, Green Gorge, Waterfall Bay and Hurd Point. The albatross researchers live at the southern end of the island, some 34 kilometres away from the main station. Occasionally, researchers will take a trip on inflatable rubber duckies—known as zodiacs—from the main station to the south of the island. However, due to the fairly consistent inclement weather, such trips are not common and

ABOVE LEFT:
Davis Point field hut was a converted water tank. It was removed from the island in 2003.

ABOVE RIGHT:
Albatrosses were handled very differently in the early days to the way they are today. Dr Ingram, an early expeditioner, holding a young albatross captured at Crozet Island. Frank Hurley, 1885–1962. National Library of Australia.

LEFT:
A researcher's hand beside a Wandering Albatross footprint gives an idea of the size of its foot.

LIVING AND WORKING ON THE ISLANDS   159

The original hut of the albatross project workers at Caroline Cove was dismantled after a landslide in 2000 and has never been rebuilt.

the main mode of transport between the huts and the main station is walking.

Living in the field is very different from living on station. The field huts are basic but comfortable. The gas heaters and stoves serve as reasonable protection against the cold. Two of the field huts are made of fibreglass and resemble big orange eggs lying on their sides. The others are made out of wood and metal and are regularly maintained by the tradespeople who work on Macquarie Island. Other field huts used in the past (now removed) have been made from fitted-out water tanks or small fibreglass huts known as 'apples'. The huts are usually resupplied once a year by helicopter. To minimise the introduction of plant diseases and pests, fresh food is not taken out into the field. Cooking ingredients therefore are generally limited to tinned and dried foods so people living in the field generally learn to become inventive cooks.

The original hut used by the albatross researchers was located at Caroline Cove in the south-west corner of the island. Named for the ship that was wrecked there in 1825, Caroline Cove is nestled under the imposing Petrel Peak and the Caroline Cove Amphitheatre, where most of the albatrosses live and breed. Unfortunately this hut was hit by a major landslide that came down the slopes of Petrel Peak in 2000 and, following this, it was deemed to be too dangerous to live and work there any more. The hut was dismantled in 2001 and albatross researchers since this time have lived at Hurd Point, some 4 kilometres away on the south-east corner.

An albatross worker generally covers 10–20 kilometres a day, and some days up to 40 kilometres. A normal day consists of walking up the steep slope out of Hurd Point to the plateau, a climb of about 200 metres, followed by a trek across the plateau to Petrel Peak, then up and over onto the southern slopes of Petrel Peak to conduct nest checks of the Black-browed and Grey-headed Albatrosses breeding there. Following these checks, it's back up and over Petrel Peak and around the amphitheatre to check the Wandering Albatross nests before eventually heading back over to Hurd Point to end the day by running down the huge gravel scree that forms a most impressive backdrop to the hut. In addition to the kilometres of horizontal distance covered, such a day also involves climbing 1–2 kilometres of altitude and albatross workers tend to become fit fairly quickly doing this type of work.

The life of a field worker is a simple one and for many of us who live and work down there on a regular basis, the solitude, simplicity and lack of distractions that dominate life in western society make it a hard lifestyle to beat. However, solitude and simplicity aside, the main driving force behind most albatross fieldworkers, indeed most people involved in this type of work, is a desire to contribute to the conservation of albatrosses, a group of magnificent seabirds that have suffered far too much from the actions of people.

A member of Sir Douglas Mawson's early expedition makes camp at the site of the Caroline Cove hut. 1911–14. State Library of New South Wales.

LIVING AND WORKING ON THE ISLANDS 161

Two Black-browed Albatross nests close together on the southern slopes of Petrel Peak, Macquarie Island.

REFERENCES

INTRODUCTION

1. **Robertson, C.J.R. and G.B. Nunn (1998).** Towards a new taxonomy for albatrosses. In *Albatross Biology and Conservation.* G. Robertson and R. Gales (eds). Surrey Beatty and Sons: 13–19. Sydney.

2. **Gales, R. (1998).** Albatross populations: status and threats. In: *Albatross Biology and Conservation.* G. Robertson and R. Gales (eds). Surrey Beatty and Sons: 20–45. Sydney.

3. **Brothers, N.B., D. Pemberton, V. Halley and H. Pryor (2001).** *Tasmania's Offshore Islands: Seabirds and Other Natural Features.* Tasmanian Museum and Art Gallery. Hobart.

4. **Warham, J. (1990).** *The Petrels: Their Ecology and Breeding Systems.* Academic Press. London.

5. **Lack, D. (1968).** *Ecological Adaptations for Breeding in Birds.* Methuen and Co. London.

6. **Gales, R. (1993).** *Co-operative Mechanisms for the Conservation of Albatross.* Parks and Wildlife Tasmania. Hobart.

7. **Croxall, J.P. (1979).** Distribution and population changes in the Wandering Albatross *Diomedea exulans* at South Georgia. *Ardea* 67: 15–21.

8. **Tomkins, R.J. (1985).** Reproduction and mortality of Wandering Albatrosses on Macquarie Island. *Emu* 85(1): 40–42.

9. **Jouventin, P., J.C. Stahl, H. Weimerskirch and J.L. Mougin (1984).** The seabirds of the French subantarctic islands and Adelie Land, their status and conservation. In *Status and Conservation of the World's Seabirds.* J.P. Croxall, P.G.H. Evans and R.W. Schreiber (eds). I.C.B.P. Technical Publication. 2. Cambridge.

10. **Weimerskirch, H. and P. Jouventin (1987).** Population dynamics of the Wandering Albatross, *Diomedea exulans*, of the Crozet Islands: causes and consequences of the population decline. *Oikos* 49(3): 315–22.

11. **Brothers, N.P., J. Cooper and S. Loekkeborg (1999).** The incidental catch of seabirds by longline fisheries: worldwide review and technical guidelines for mitigation. Food and Agricultural Organisation (FAO-UN) *Fisheries Circular No. 937.* Rome.

12. **Croxall, J.P., P.A. Prince, P. Rothery and A.G. Wood (1998).** Population changes in albatrosses at South Georgia. In *Albatross Biology and Conservation.* G. Robertson and R. Gales (eds). Surrey Beatty and Sons: 69–83. Sydney.

13. **Weimerskirch, H. and P. Jouventin (1998).** Changes in population sizes and demographic parameters of six albatross species breeding on the French sub-Antarctic islands. In *Albatross Biology and Conservation.* G. Robertson and R. Gales (eds). Surrey Beatty and Sons: 84–91. Sydney.

14. **Inchausti, P. and H. Weimerskirch (2001).** Risks of decline and extinction of the endangered Amsterdam Albatross and the projected impact of long-line fisheries. *Biological Conservation* 100: 377–86.

15. **Nel, D.C., P.G. Ryan, J.M. Crawford, J. Cooper and O. Huyser (2002).** Population trends of albatrosses and petrels at sub-antarctic Marion Island. *Polar Biology* 25: 81–89.

16. **Waugh, S.M., H. Weimerskirch, P.J. Moore and P.M. Sagar (1999).** Population dynamics of Black-browed and Grey-headed Albatrosses *Diomedea melanophrys* and *D. chrysostoma* at Campbell Island, New Zealand, 1942–96. *Ibis* 141(2): 216–25.

17. **de la Mare, W.K. and K.R. Kerry (1994).** Population dynamics of the Wandering Albatross (*Diomedea exulans*) on Macquarie Island and the effects of mortality from longline fishing. *Polar Biology* 14(4): 231–41.

18. **Tuck, G., T. Polacheck and C. Bulman (2003).** Spatio-temporal trends in long-line fisheries of the Southern Ocean and implications for seabird by-catch. *Biological Conservation* 114: 1–27.

19. **Brothers, N. (1991)** Albatross mortality and associated bait loss in the Japanese long line fishery in the Southern Ocean. *Biological Conservation* 55: 255–68.

20. **Nel, D.C., P.G. Ryan and B.P. Watkins (2002).** Seabird mortality in the Patagonian Toothfish longline fishery around the Prince Edward Islands, 1996–2000. *Antarctic Science* 14(2): 151–61.

BREEDING AREAS

1. **Rintoul, S. (2000).** Southern Ocean currents and climate. *Papers and Proceedings of the Royal Society of Tasmania.* **133**(3): 41–50.

2. **Rodhouse, P.G. (1996).** Cephalopods and mesoscale oceanography at the Antarctic Polar Front: satellite tracked predators locate pelagic trophic interactions. *Marine Ecology Progress Series* **136**: 37–50.

3. **Waugh, S.M., H. Weimerskirch, Y. Cherel, U. Shankar, P.A. Prince and P.M. Sagar (1999).** Exploitation of the marine environment by two sympatric albatrosses in the Pacific Southern Ocean. *Marine Ecology Progress Series* **177**: 243–54.

4. **Nel, D.C., J.R.E. Lutjeharms, E.A. Pakhomov, I.J. Ansorge, P.G. Ryan and N.T.W. Klages (2001).** Exploitation of mesoscale oceanographic features by Grey-headed Albatross *Thalassarche chrysostoma* in the southern Indian Ocean. *Marine Ecology Progress Series* **217**: 15–26.

5. **Ashmole, N.P. (1971).** Seabird Ecology and the Marine Environment. In *Avian Biology.* D.S. Farner and J.R. King (eds). Academic Press. **1**: 223–86. London.

6. **Jouventin, P. and H. Weimerskirch (1990).** Satellite tracking of Wandering Albatrosses. *Nature* **434**: 746–48.

7. **Selkirk, P.M., R.D. Seppelt and D.R. Selkirk (1990).** *Subantarctic Macquarie Island: Environment and Biology.* Cambridge University Press. Melbourne.

8. **Gordon, A.L., H.W. Taylor and D.T. Georgi (1977).** Antarctic oceanographic zonation. In *Polar Oceans.* M.J. Dunbar (ed.) Arctic Institute of North America: 45–76. Calgary.

9. **Belkin, G. (1996).** Southern Ocean Fronts from the Greenwich Meridian to Tasmania. *Journal for Geophysical Research* **101**: 3675–96.

10. **Trull, T., S.R. Rintoul, M. Hadfield and E.R. Abraham (2001).** Circulation and seasonal evolution of polar waters south of Australia: Implications for iron fertilization of the Southern Ocean. *Deep-Sea Research Part II- Topical Studies in Oceanography* **48**(11–12): 2439–66.

11. **Orsi, A.H., T. Whitworth and W.D. Nowlin (1995).** On the meridional extent and fronts of the antarctic circumpolar current. *Deep-Sea Research Part I: Oceanographic Research Papers* **42**(5): 641–73.

12. **Piontkovski, S. A., R. Williams and T.A. Malnik (1995).** Spatial heterogeneity biomass and size structure of plankton of the Indian Ocean: some general trends. *Marine Ecology Progress Series* **117**: 219–27.

13. **Hunt, B.P.V., E.A. Pakhomov and C.D. McQuaid (2001).** Short-term variation and long-term changes in the oceanographic environment and zooplankton community in the vicinity of a sub-Antarctic archipelago. *Marine Biology* **138**(2): 369–81.

14. **Brothers, N. (personal communication).** Observations on Bishop and Clerk Islets, 1993.

15. **Copson, G. R. (1988).** Status of the Black-bowed and Grey-headed Albatrosses on Macquarie Island. *Proceedings of the Royal Society of Tasmania* **122**(1): 137–41.

16. **Csordas, S.E. and R. Carrick (1965).** The New Zealand Fur Seal, *Arctocephalus forsteri* (Lesson), at Macquarie Island, 1949–1964. *CSIRO Wildlife Research* **10**: 83–99.

17. **Goldsworthy, S.D., L. Wynen, S. Robinson and P. Shaughnessy (1998).** The population status and hybridisation of three sympatric fur seals (*Arctocephalus* spp.) at Macquarie Island. *New Zealand Natural Sciences, Supplement* **23**: 68.

18. **McMahon C.R., M.N. Bester, H.R. Burton, M.A. Hindell and C. J. A. Bradshaw (2005).** Population status, trends and a re-examination of the hypotheses explaining the recent declines of the southern elephant seal *Mirounga leonina*. *Mammal Review* **35**: 82–100.

19. **Shaw, J.D. (2005).** *Reproductive Ecology of Subantarctic Vegetation on Macquarie Island.* School of Botany. PhD Thesis, University of Tasmania, Hobart.

20. **Brothers, N. *et al*. (2001).** Tasmanian Museum and Art Gallery. Hobart.

21. **Dixon, G. (1996).** *A Reconnaissance Inventory of Sites of Geoconservation Significance on*

22. **Kostoglou, P.** (**1996**). *Sealing in Tasmania Historical Research Project*. A Report for the Parks and Wildlife Service Tasmania. Hobart.

Tasmanian Islands. Earth Sciences Section, Department of Environment and Land Management. Hobart.

23. **Banks, M.** (**1993**). *Reconnaissance Geology and Geomorphology of the Major Islands South of Tasmania*. Department of Parks, Wildlife and Heritage. Hobart.

24. **Brothers, N., A. Wiltshire, D. Pemberton, N. Mooney and B. Green** (**2003**). The feeding ecology and field energetics of the Pedra Branca skink *Niveoscincus palfreymani*. *Wildlife Research* **30**: 81–87.

25. The bathymetry layer was created using the ETOPO2 database provided by National Geophysical Data Centre (NGDC-http://www.ngdc.noaa.gov/mgg/fliers/01mgg04.html). The primary source of seafloor data is Smith, W.H.F. and D.T. Sandwell (1997). Global seafloor topography from satellite altimetry and ship depth soundings. *Science* **277**: 1957–62.

26. The AVHRR Oceans Pathfinder Sea Surface temperature data was obtained from the Physical Oceanography Distributed Active Archive Center (PO.DAAC) at the NASA Jet Propulsion Laboratory, Pasadena, CA. (http://podaac.jpl.nasa.gov).

27. Sea surface height anomaly data was downloaded from World Ocean Experiment (WOCE3). Sea Level from Topex/Poseidon satellite courtesy of WOCE Data Products Committee 2002; WOCE Global Data: Satellite Data, Version 3.0; WOCE International Project Office, WOCE Report No. 180/02, Southampton, UK.

HUMANS IN BASS STRAIT AND THE SOUTHERN OCEAN

1. **Hughes, R.** (**1988**). *The Fatal Shore—The Epic of Australia's Founding*. Vintage Books. Random House. New York.

2. **Nash, M.** (**2003**). *The Bay Whalers: Tasmania's Shore Based Whaling Industry*. Navarine Publishing. Hobart.

3. **Nicholls, M. (ed.)** (**1977**). *The Diary of the Reverend Robert Knopwood 1803–1838*. Tasmanian Historical Research Association. Hobart.

4. **Dawbin, W.** (**1986**). Right Whales caught in waters around south eastern Australia during the 19th and early 20th century. *Reports of the International Whaling Commission* Special Issue **10**: 261–67.

5. **Kostoglou, P.** (**1996**). Parks and Wildlife Service Tasmania. Hobart.

6. **Ling, J.K.** (**1999**). Exploitation of Fur Seals and Sea Lions from Australia, New Zealand and adjacent subantarctic islands during the 18th, 19th and 20th centuries. *Australian Zoologist* **31**: 323–50.

7. **Cumpston, J.S.** (**1968**). *Macquarie Island. ANARE Scientific Reports, Series A(1) Narrative*. Antarctic Division, Department of External Affairs. Melbourne.

8. **Shaughnessy, P.D. and S.D. Goldsworthy** (**1993**). Feeding ecology of Fur Seals and their management at Heard and Macquarie Islands. Proceedings NIPR Symposium. *Polar Biology* **6**: 173–75.

9. **Shaughnessy, P.D., G.L. Shaughnessy and L. Fletcher** (**1988**). Recovery of the Fur Seal population at Macquarie Island. *Papers and Proceedings of the Royal Society of Tasmania* **122(1)**: 177–87.

10. **Plomley, N.J.B.E.** (**1966**). *Friendly Mission. The Tasmanian journals and papers of George Augustus Robinson*. The Tasmanian Historic Research Association. Hobart.

11. **Backhouse, J.** (**1843**). *A narrative of a visit to the Australian colonies*. Hamilton, Adams and Co. London.

12. **Ashworth, H.P.C. and D. le Souef** (**1895**). Albatross Island and the Hunter Group. *The Victorian Naturalist* **11**: 134–44.

13. **Flinders, M.** (**1814**). *A Voyage to Terra Australis*. G. & W. Nicol. London.

14. **Tuck, G. *et al.*** (**2003**). *Biological Conservation* **114**: 1–27.

15. **Gales, R. (1998).** In *Albatross Biology and Conservation*. G. Robertson and R. Gales (eds). Surrey Beatty and Sons: 20–45. Sydney.

16. **Brothers, N. (1991).** *Biological Conservation* 55: 255–68.

17. **Brothers, N. et al. (1999).** FAO *Fisheries Circular No. 937*. Rome.

18. **de la Mare, W.K. and K.R. Kerry (1994).** *Polar Biology* 14 (4): 231–41.

19. **Croxall, J.P. et al. (1998).** In *Albatross Biology and Conservation*. G. Robertson and R. Gales (eds). Surrey Beatty and Sons: 69–83. Sydney.

20. **Gales, R., N. Brothers and T. Reid (1998).** Seabird mortality in the Japanese tuna longline fishery around Australia, 1988–1995. *Biological Conservation* 86(1): 37–56.

21. **Murray, T.E., J.A. Bartle, S.R. Kalish and P.R. Taylor (1993).** Incidental capture of seabirds by Japanese southern bluefin tuna long line vessels in New Zealand waters, 1988–1992. *Bird Conservation International* 3: 181–210.

22. **Brothers, N., R. Gales, A. Hedd and G. Robertson (1998).** Foraging movements of the Shy Albatross *Diomedea cauta* breeding in Australia—Implications for interactions with longline fisheries. *Ibis* 140(3): 446–57.

23. **Dalziell, J. and M.D. Poorter (1993).** Seabird mortality in longline fisheries around South Georgia. *Polar Record* 29(169): 143–45.

24. **Nel, D.C., P.G. Ryan, J.L. Nel, N.T.W. Klages, R.P. Wilson, G. Robertson and G.N. Tuck (2002).** Foraging interactions between Wandering Albatrosses *Diomedea exulans* breeding on Marion Island and long-line fisheries in the southern Indian Ocean. *Ibis* 144(3): E141–E154.

25. **Ryan, P.G. and C. Boix Hinzen (1999).** Consistent male-biased seabird mortality in the Patagonian Toothfish longline fishery. *Auk* 116(3): 851–54.

26. **Ryan, J.P., C. Boix-Hinzen, J.W. Enticott, D.C. Nel, R. Wanless and M. Purves (1997).** Seabird mortality in the long-line fishery for Patagonian Toothfish at the Prince Edward Islands: 1996–1997. Hobart, CCAMLR, WGA-FSA-97/51: 15. Cited in Brothers *et al.* (1999).

27. **Baker, G.B., R. Gales, S. Hamilton and V. Wilkinson (2002).** Albatrosses and Petrels in Australia: a review of their conservation and management. *Emu* 102: 71–97.

28. **Weineke, B. and G. Roberston (2002).** Seabird and seal fisheries interactions in the Australian Patagonian Toothfish *Dissostichus eleginoides* trawl fishery. *Fisheries Research* 54: 253–65.

THE ALBATROSSES

1. **Marchant, S. and P.J. Higgins (1990).** *Handbook of Australian, New Zealand and Antarctic Birds*. Royal Australian Ornithologists Union. Melbourne.

2. **Tomkins, R.J. (1984).** Some aspects of the morphology of Wandering Albatrosses on Macquarie Island. *Emu* 84(1): 29–32.

3. **Terauds, A. and R. Gales (in preparation).** Does variation in demographic and environmental parameters influence breeding success of albatrosses on Macquarie Island? *Manuscript in Preparation*.

4. **Weimerskirch, H. and P. Lys (2000).** Seasonal changes in the provisioning behaviour and mass of male and female Wandering Albatrosses in relation to the growth of their chick. *Polar Biology* 23(11): 733–44.

5. **Terauds, A., R. Gales, R. Alderman and G.B. Baker (in preparation).** Population and survival trends of Wandering Albatrosses *Diomedea exulans* breeding on Macquarie Island. *Manuscript in preparation*.

6. **Carrick, R. and S.E. Ingham (1970).** Ecology and population dynamics of Antarctic seabirds. In *Antarctic Ecology*. M.W. Holdgate (ed.). Academic Press: 502–25. New York.

7. **de la Mare, W. K. and K. R. Kerry (1994).** *Polar Biology* 14(4): 231–41.

8. **Jouventin, P. and H. Weimerskirch (1990).** *Nature* 434: 746–48.

9. **Weimerskirch, H. and P. Jouventin (1998).** In *Albatross Biology and Conservation*. G. Robertson and R. Gales (eds). Surrey Beatty and Sons: 84–91. Sydney.

A Light-mantled Sooty Albatross soars high above the coastal slopes of Macquarie Island.

10. **Prince, P.A., J.P. Croxall, P.N. Trathan and A.G. Wood** (1998). The pelagic distribution of South Georgia Albatrosses and their relationships with fisheries. In *Albatross Biology and Conservation*. G. Robertson and R. Gales (eds). Surrey Beatty and Sons: 137–67. Sydney.

11. **Rodhouse, P. G.** (1996). *Marine Ecology Progress Series* 136: 37–50.

12. **Croxall, J.P. and P.A. Prince** (1994). Dead or alive, night or day—How do albatrosses catch squid? *Antarctic Science* 6(2): 155–62.

13. **Ryan, P.G. and C. Boix Hinzen** (1999). *Auk* 116(3): 851–54.

14. **Brothers, N.P.** *et al.* (1998). *Ibis* 140(3): 446–57.

15. **Hedd, A., R. Gales and N. Brothers** (2001). Foraging strategies of Shy Albatross *Thalassarche cauta* breeding at Albatross Island, Tasmania, Australia. *Marine Ecology Progress Series* 224: 267–82.

16. **Hedd, A., R. Gales and N. Brothers** (2002). Provisioning and growth rates of Shy Albatrosses (*Thalassarche cauta*) breeding in Australia. *The Condor* 104: 12–29.

17. **Brothers, N.P., T.A. Reid and R.P. Gales** (1997). At-sea distribution of Shy Albatrosses *Diomedea cauta cauta* derived from records of band recoveries and colour-marked birds. *Emu* 97: 231–39.

18. **Brothers, N.P.** *et al.* (2001). Tasmanian Museum and Art Gallery. Hobart.

19. **Gales, R.** *et al.* (1998). *Biological Conservation* 86(1): 37–56.

20. **Hedd, A. and R. Gales** (2001). The diet of Shy Albatrosses (*Thalassarche cauta*) at Albatross Island, Tasmania. *Journal of Zoology* 253: 69–90.

21. **Hedd, A., R. Gales, N. Brothers and G. Robertson** (1997). Diving behaviour of the Shy Albatross *Diomedea cauta* in Tasmania—Initial findings and dive recorder assessment. *Ibis* 139: 452–60.

22. **Brothers, N.** (personal communication) Observations on Bishop and Clerk Islets, 1993.

23. **Terauds, A., R. Gales and R. Alderman** (in press). Trends in population numbers and survival of Black-browed and Grey-headed Albatrosses on Macquarie Island. *Emu*.

24. **Terauds, A. and R. Gales** (in preparation). Breeding frequency of albatrosses on Macquarie Island. *Manuscript in preparation*.

25. **Copson, G.R.** (1988). *Proceedings of the Royal Society of Tasmania* 122(1): 137–41.

26. **Mackenzie, D.** (1968). The birds and seals of the Bishop and Clerk Islets, Macquarie Island. *Emu* 67(4): 240–45.

27. **Lugg, D.J., G.W. Johnstone and B.J. Griffin** (1978). The outlying islands of Macquarie Island. *The Geographical Journal* 144-2: 277–88.

28. **Robinson, S.A., S.G. Goldsworthy, J. van den Hoff and M. A. Hindell** (2002). The foraging ecology of two sympatric Fur Seal species, *Arctocephalus gazella* and *Arctocephalus tropicalis*, at Macquarie Island during the austral summer. *Marine and Freshwater Research* 53(7): 1071–82.

29. **Terauds, A., R. Gales, G.B. Baker and R. Alderman** (in press). Foraging areas of Black-browed and Grey-headed Albatrosses from Macquarie Island in relation to marine protected areas. *Aquatic Conservation—Marine and Freshwater Ecosystems*.

30. **Nel, D.C.** *et al.* (2001). *Marine Ecology Progress Series* 217: 15–26.

31. **Terauds, A. and R. Gales** (in preparation). Provisioning strategies of Light-mantled Sooty Albatrosses on Macquarie Island. *Manuscript in Preparation*.

32. **Weimerskirch, H. and G. Robertson** (1994). Satellite Tracking of Light-mantled Sooty Albatrosses. *Polar Biology* 14(2): 123–26.

33. **Weimerskirch, H., O. Chastel, L. Ackermann, T. Chaurand, F. Cuenotchaillet, X. Hindermeyer and J. Judas** (1994). Alternate long and short foraging trips in pelagic seabird parents. *Animal Behaviour* 47(2): 472–76.

34. **The bathymetry layer was created using the ETOPO2 database provided by National Geophysical Data Centre** (NGDC-http://www.ngdc.noaa.gov/mgg/fliers/01mgg04.html). Smith, W. H. F. and D. T. Sandwell *Science*, 277: 1957–62.

CONSERVATION

1. **Weimerskirch, H. and P. Jouventin (1998).** In *Albatross Biology and Conservation.* G. Robertson and R. Gales (eds). Surrey Beatty and Sons: 84–91. Sydney.

2. **Croxall, J.P. *et al.* (1998).** In *Albatross Biology and Conservation.* G. Robertson and R. Gales (eds). Surrey Beatty and Sons: 69–83. Sydney.

3. **Walker, K. and G. Elliott (1999).** Population changes and biology of the Wandering Albatross *Diomedea exulans gibsoni* at the Auckland Islands. *Emu* 99(4): 239–47.

4. **Waugh, S.M., H. Weimerskirch, P.J. Moore and P.M. Sagar (1999).** Population dynamics of Black-browed and Grey-headed Albatrosses *Diomedea melanophrys* and *D. chrysostoma* at Campbell Island, New Zealand, 1942–96. *Ibis* 141(2): 216–25.

5. **Stahl, J.C. and P.M. Sagar (2000).** Foraging strategies of southern Buller's Albatrosses *Diomedea b. bulleri* breeding on The Snares, New Zealand. *Journal of the Royal Society of New Zealand* 30(3): 299–318.

6. **Nel, D.C. *et al.* (2000).** *Biological Conservation* 96(2): 219–31.

7. **Nel, D.C. *et al.* (2002).** *Polar Biology* 25: 81–9.

8. **Inchausti, P. and H. Weimerskirch (2001).** Risks of decline and extinction of the endangered Amsterdam Albatross and the projected impact of long-line fisheries. *Biological Conservation* 100: 377–86.

9. **Macdonald, D. and R.H. Green (1963).** Albatross Island. *Emu* 63: 23–31.

10. **Johnstone, G.W., D. Milledge and D.F. Dorward (1975).** The White-capped Albatross of Albatross Island: numbers and breeding behaviour. *Emu* 75: 1–11.

11. **Gales, R. (1993).** Parks and Wildlife Tasmania. Hobart.

12. **Brothers, N. (1991).** *Biological Conservation* 55: 255–68.

13. **Brothers, N. *et al.* (1998).** *Ibis* 140(3): 446–57.

14. **Hedd, A. *et al.* (2001).** *Marine Ecology Progress Series* 224: 267–82.

15. **Gales, R. *et al.* (1998).** *Biological Conservation* 86(1): 37-56.

16. **Brothers, N.P. *et al.* (1997).** *Emu* 97: 231–39.

17. **Selkirk, P. *et al.* (1990)** Cambridge University Press. Melbourne.

18. **de la Mare, W.K. and K.R. Kerry (1994).** *Polar Biology* 14(4): 231–41.

19. **Copson, G.R. (1988).** *Proceedings of the Royal Society of Tasmania* 122(1): 137–41.

20. **Croxall, J.P. and R. Gales (1998).** An assessment of the conservation status of albatrosses. In *Albatross Biology and Conservation.* G. Robertson and R. Gales (eds). Surrey Beatty and Sons: 46–66. Sydney.

The pale, guano-covered island of Pedra Branca stands out against the overcast sky of southern Tasmania.

A four-week old Wandering Albatross chick bathed in afternoon sunlight on Petrel Peak, Macquarie Island.

Artists and Photographers

ARTISTS

Fiona Stewart
Peter Hall — *Page 81*
John Gale — *Page 80*

PHOTOGRAPHERS

Aleks Terauds
Tore Pedersen — *Page 33, 147*
Justine Shaw — *Page 53*
Graham Robertson — *Pages 54, 67, 71, 72–3*
Felicity Jenkins — *Page 62*
Rachael Alderman — *Pages 68-69*

LIBRARY AND MUSEUM IMAGES

Illustrations and photos used on the following pages have been reproduced by permission of:

National Library of Australia — *Pages 3, 52, 55, 56, 62, 81, 95, 106, 121, 159*

State Library of New South Wales — *Pages 59, 161*
State Library of Victoria — *Pages 57, 61, 63*
Queen Victoria Museum and Art Gallery — *Page 56*

The rugged coastline of the west coast of Macquarie Island. Sandell Bay, with the less prominent Davis Bay in the foreground.

A snow-covered Mt Haswell forms a stark backdrop to the Caroline Cove amphitheatre, home to many of the breeding Wandering Albatrosses on Macquarie Island.

INDEX

A

Agreement on the Conservation
of Albatrosses and Petrels (ACAP)
145, 148, 152
Alauda arvensis 36
Albatross Island
   albatross population 9
   breeding site 19
   flora and fauna 35–6
   living and working on 155–6
   location 19, 35
   management and research 36
   photographs 34, 37
   topography 35
albatrosses *see also* species names
   breeding areas 19–21
   conservation efforts 145–53
   conservation status 17, 148
   deaths 12, 67
   feeding 20
   life history 9, 21
   population decline 9, 12, 17
   research 17, 145–53
   threats to 13, 17, 54, 63, 67, 70, 147–8
Amsterdam Albatross 145
Amsterdam Island 12
ANARE 33, 150
Antarctic Circumpolar Current
18–19, 20, 23–4
Aptenodytes patagonicus 25
Arctocephalus
   forsteri 29
   gazella 29
   pusillus doriferus 36
   tropicalis 29
Arenaria interpres 36
Asplenium obstusatum 50
Azorella macquariensis 29
Australian National Antarctic Research
Expedition (ANARE) 33, 150

B

banding, bird 146, 150
Big Bird Race 148

Black-browed Albatross
   bill shape 76
   breeding cycle 13, 105–11
   breeding populations 9
   breeding sites 19, 21, 24–5, 105
   features 75
   feeding habits 115–16
   flight profile 77
   life history 105–11
   photographs and illustrations
72–4, 76–7, 104–17, 162
   population trends 111, 115
Black-footed Albatross
   breeding sites 19
   deaths 12
Blackbird, Common 36

C

Cabbage, Macquarie Island 29, 31, 33
Carduleis flammea 28
Carpobrotus rossii 50
Catharacta lonbergi 28
cats, feral 28, 64
Circumpolar Current 18–19, 20, 23–4
Circus approximans 36
Colobanthus muscoides 29
Cormorant 28
   Black-faced 44, 50
Corvus tasmanicus 36
currents, ocean 18–19, 23–4
Cushion Plant 29
   Coastal 29

D

Diomedea
   amsterdamensis 145
   exulans 9, 79
Disphyma crassifolium 36
Dissostichus eleginoides 13, 53, 70
driftnet fisheries 12
ducks 28
Durvillaea antarctica 30

E

Eubalaena australis 57
Eudyptes
   chrysocome filholi 25
   schlegeli 25
Eudyptula minor 36
explorers 53

F

Falco berigora 36
Falcon, Brown 36
Falkland Islands 19
Fantail, Grey 36
feral animals 28, 33, 64–6
Fern, Macquarie Island 27
ferns 33
fisheries
   driftnet 12
   long-line 12, 53–4, 66–7, 70, 147–8
   trawling 71

G

Galapagos Islands 19
Gannet, Australasian 41–4
giant petrel 70, 72–3, 84, 151
Glasswort, Beaded 44
grasses 33
Grey-headed Albatross
   bill shape 76
   breeding cycle 13, 119–26
   breeding populations 9
   breeding sites 19, 21, 24–5, 119
   features 75
   feeding habits 130
   flight profile 77
   life history 119–26
   photographs and illustrations
12, 13, 21, 74, 76–7, 118–31
   population trends 126
Gull
   Kelp 28, 44
   Pacific 36, 44
   Silver 36, 44, 50

H
Haemotopus fuliginosus 36
Haliaeetus leucogaster 36
Halobeana caerulea 28
Harrier, Swamp 36
Hawaiian islands 19
Heard Island 19

I
Iles Crozet 12
Iles Kerguelen 12, 19
International Biosphere Reserve 33

K
kelp 30

L
Larus
    dominicanus 28
    novaehollandiae 36
    pacificus 36
Laysan Albatross
    breeding sites 19
    deaths 12
Light-mantled Sooty Albatross
    bill shape 76
    breeding cycle 13, 133–9
    breeding populations 9
    breeding sites 19, 21, 25, 133
    features 75
    feeding habits 142
    flight profile 77
    life history 133–9
    photographs and illustrations
        14–15, 17, 74, 76–7, 132–44, 167
    population trends 142
long-line fisheries
    12, 53–4, 66–7, 70, 147–8

M
Macquarie Island
    albatross population 9, 12
    feral animals 64–6
    flora and fauna 24–32
    living and working on 157–61
    location 20, 23
    management and research 17, 33, 150–2
    map 23
    photographs 24–32, 171
    topography 23
Macronectes
    giganteus 25
    halli 25
Marion Island 12, 19
megaherbs 26, 29, 33
Mewstone
    albatross population 9
    breeding site 19, 50
    flora and fauna 50
    living and working on 156–7
    location 19, 47
    management and research 50
    photographs 46–51
    topography 47
Mirounga leonina 28
muttonbirds 63

N
Nematoceras dienema 33
Neophoca cinerea 57
New Zealand 12
Niveoscincus
    metallica 36
    palfreymani 41
    pretiosa 36, 50

O
orchids 33
Oystercatcher, Sooty 36

P
Pachyptila
    desolata 28
    turtur 36
Patagonian Toothfish fisheries
    13, 53, 70–1

Pedra Branca
    albatross population 9
    breeding site 19, 41, 44
    flora and fauna 41, 44
    living and working on 156
    location 19, 41
    management and research 44
    photographs 40–1
    Skink 41, 44, 45
    topography 41
Penguin
    Gentoo 25, 26
    King 25, 32, 64
    Little 36
    Rockhopper 25, 32
    Royal 22, 25, 30, 64
Petrel
    Blue 28
    giant 70, 72–3, 84, 151
    Grey 28, 70
    Northern Giant 25, 28, 31
    Southern Giant 25, 28, 32
    White-chinned 70
    White-headed 28
petrels, threats to 54, 70
Phalacrocorax
    albiventer purpurascens 28
    fuscescens 44
Phocarctos hookeri 29
Phoebastria
    albatrus 12
    immutabilis 9
    irrorata 9
    nigripes 9
Phoebetria
    fusca 75
    palpebrata 9, 133
Physeter macrocephalus 57
Pigface 50
Pleurophyllum hookeri 26, 29
Poa
    foliosa 29
    poiformis 36

Polystichum vestitum 27, 33
Prion
    Antarctic 28
    Fairy 36, 50
Procellaria
    aequinoctialis 70
    cinerea 28
Pterodroma lessonii 28
Puffinus
    griseus 28
    tenuirostris 36
Pygoscelis papua papua 25

R

rabbits 28, 33, 64–6
Raven, Forest 36
Redpolls 28
Rhipidura fuliginosa 36

S

Salicornia quinqueflora 44, 50
satellite-tracking 20, 145, 147–50
Sea-Eagle, White-bellied 36
Sea Lion
    Australian 57, 59
    New Zealand 29
Seabirds 28, 63–4
Seal
    Antarctic Fur 29
    Australian Fur 36, 44, 50, 57, 58, 59
    Elephant 52, 58, 61
    New Zealand Fur 29, 36, 57, 59, 60
    Southern Elephant
        22, 26, 28–9, 57, 58, 59
    Subantarctic Fur 29, 52, 59
sealing
    Macquarie Island 59–63
    Tasmania 57–9
Senecio lautus 50
Senecio sp. 36
settlers, early 53, 57
Shag, Blue-eyed 27, 28

Shearwater
    Short-tailed 36, 63
    Sooty 28
Short-tailed Albatross
    breeding sites 19
    deaths 12
Shy Albatross
    bill shape 76
    breeding cycle 13, 93–9
    breeding populations 9
    breeding sites 19, 21, 35, 41, 50, 93
    features 75
    feeding habits 102
    flight profile 77
    life history 93–9
    photographs and illustrations
        8, 16, 34, 38–40, 46, 49–51, 74,
        76–7, 92–103, 148
    population trends 99
Silvereye 36
Skink
    Metallic 36
    Pedra Branca 44–5
    Tasmanian Tree 36, 50
Skua, Subantarctic 28
Skylark 36
South Georgia 9, 12, 19
Southern Ocean, breeding sites 18
Starling, Common 28, 36
Sterna vittata bethunei 28
Stilbocarpa polaris 29
Sturnus vulgaris 28, 36
Sula serrator 41

T

Tasmanian Department of Primary
    Industries, Water and Environment
    (DPIWE) 44, 50, 66
Tasmanian Parks and Wildlife Service
    33, 50, 66
Tern, Antarctic 28

Thalassarche
    cauta 9, 93
    chrysostoma 9, 119
    melanophrys 9, 105
tracking, bird *see* satellite-tracking
trawling 71
tuna fisheries 12, 53, 66–7, 70
Turdus merula 36
Turnstone, Ruddy 36
Tussock Grass 26, 29, 31, 33
    Blue 36, 50

W

Wandering Albatross
    bill shape 9, 76
    breeding cycle 13, 79–90
    breeding populations 9
    breeding sites 19, 21, 24, 79
    features 75
    feeding habits 91
    flight profile 77
    life history 79–90
    photographs and illustrations
        9, 10–11, 13, 21, 68–9, 72–4,
        76–91, 145, 170
    population trends 91
water temperatures 24, 35, 36
Waved Albatross 19
wekas 64
Whales
    Southern Right 57
    Sperm 57
whaling 55–7
wildlife, exploitation 53–64
World Heritage Area 33, 44

Z

Zosterops lateralis 36

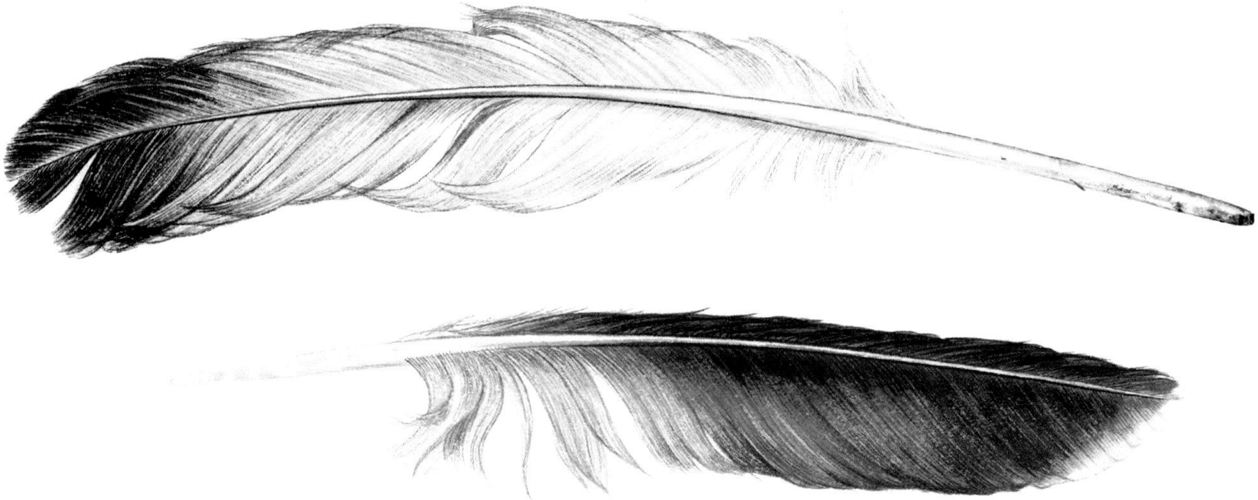

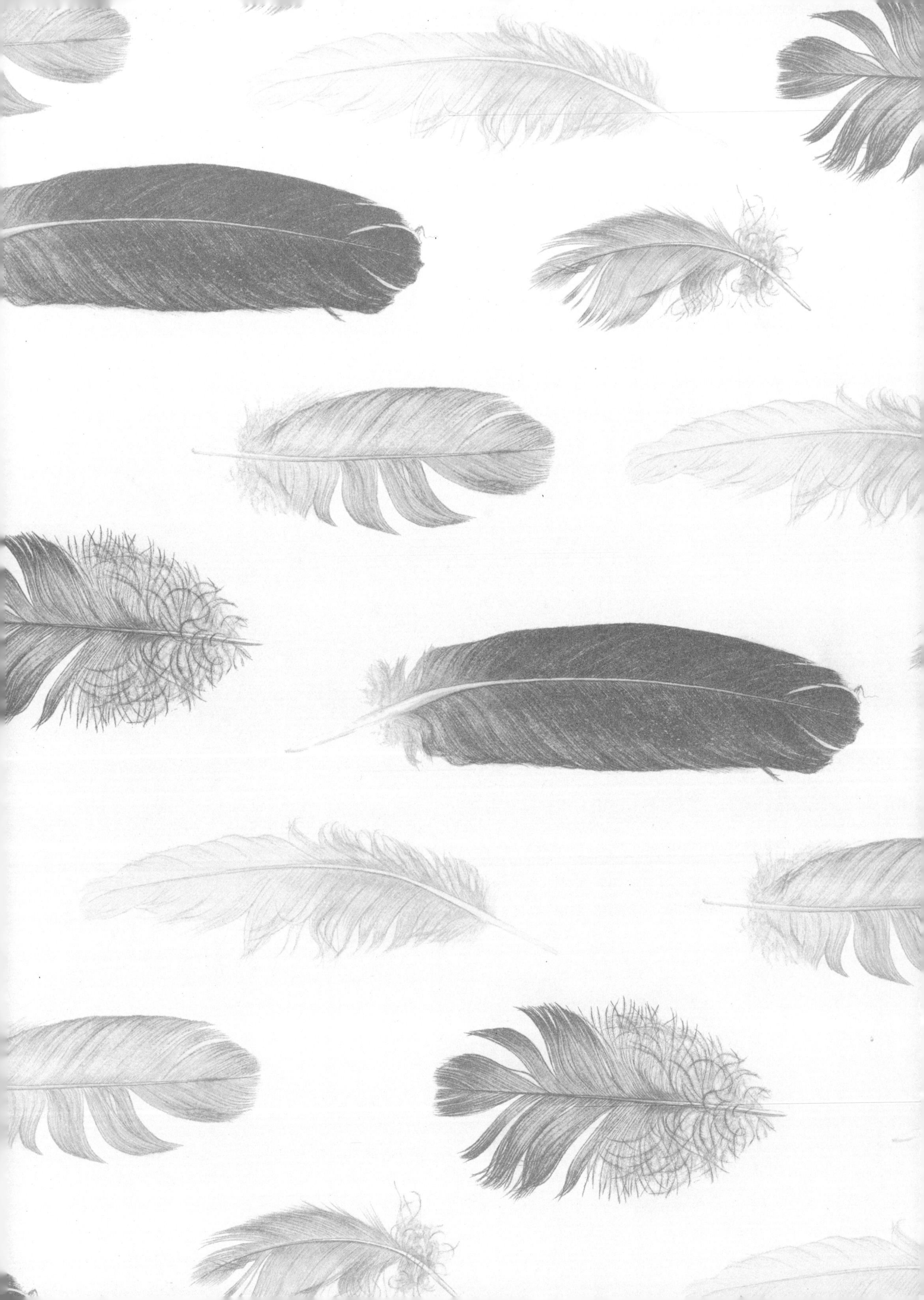